The Cybersecurity Mindset:
A Virtual and Transformational Thinking Mode

by Dewayne Hart

ISBN 978-1-64663-588-7

Published by

800 Battery Ave SE
Suite #100
Atlanta, GA 30339, USA
(470) 409-8316
www.dewaynehart.com

THE CYBER SECURITY MINDSET

A VIRTUAL AND TRANSFORMATIONAL THINKING MODE

DEWAYNE HART

SEMAIS
Cybersecurity Consulting

TABLE OF CONTENTS

INTRODUCTION

HOW DID WE GET HERE?

Today's technology has survived many milestones and challenges. In the 1980s, IBM created the first personal computer during the microcomputer revolution. Before this era, mainframe computers only supported data manipulation. The IBM Model number 5150 surfaced on August 12, 1981, and created a new technology environment. During the same period, the UK introduced the Sinclair Zx81 computer, and Microsoft channeled the market with MS-DOS as the premier operating system supporting IBM-PC compatible computers. According to Microsoft, in 1994, MS-DOS was operating on 100 million computers worldwide.

In 1995, I started my post-sea-duty career or shore-duty at MacDill AFB, which was where the experience and exposure to the PC market surfaced. From 1995 to 2000, Microsoft software products and technology controlled the IT market. Many government agencies transitioned to newer technologies that were Windows-based. The internet began to surface during this trend, and as an IT professional, my technology engagement advanced. The internet became a viable source for linking these computers and a vehicle to support data transactions and multiple communication technologies—such as cell phones, modems, and military tactical systems.

In 1995, Wells Fargo became the first US bank to offer online banking, with other banks quickly following suit. Here is where my professional career in technology and security surfaced. I remember speaking to several military friends about securing data and protection standards and how computer viruses would dominate data protection and internet safety. Since the concept was new and far from a concern, I visualized technology encountering many challenges; today, professionals are multi-challenged to defend and protect systems.

From the early 2000s to 2010, I saw many organizations develop data protection standards. This massive growth onboarded a new culture and supporting technologies, and cybersecurity became a premier concern for IT managers. Organizations integrated safe practices to protect data and monetary loss. The online banking industry exploded, and so did social media—Facebook, Instagram, and Twitter. The industry saturated the market and created a chain of protection standards, frameworks, and social-behavioral issues. The result forced technology to grasp more understanding and meaning for security.

IT SECURITY 101

The three pillars of IT security are Confidentiality, Integrity, and Availability—commonly called the CIA. Confidentiality is a principle that describes a need-to-know basis. For instance, not everyone should have access to your bank account. That's why access requires a separate username. The creation of shared accounts can break the confidentiality scheme. Integrity is defined as free from modification. That means data transmitted and received should mirror the same . format. If you transfer $1,000 to your significant other for Valentine's

Day, their account should increase by $1,000, not by $10,000. Of course, they may like the digits—but sorry for you! You cannot take it back. Here is where integrity comes active. Our last principle is availability. Availability ensures that resources are available, such as a secure communication channel when executing the banking transfer, and your passwords are encrypted. Encryption enables confidentiality. It's a secret representation of your password. When you type a password such as "SDER%$&JHV) *;jh," it is converted into a possible 1,024 character with unique codes. Let's not get too technical—but you see the point. There are various forms of availability, such as logging onto a system during specific periods. Some key areas are uptime, storage access, or accessing social media sites.

In the realm of IT, security vulnerabilities and threats exist. A vulnerability is a weakness or loophole, such as a password structure. If an organization requires employee accounts to use fifteen-character passwords, and a user can successfully create a four-digit password—that's a vulnerability. Threats exploit vulnerabilities—this would be a hacker (threat agent). The hacker could have prior knowledge of the password complexity requirements and gain access to confidential information—such as an employee email message: "I have a four-character password!"

IT systems utilize logical rules to counter the risk, such as a fifteen-character password. A hacker can use various password-guessing methodologies. One is to execute a dictionary attack by generating common dictionary words using hacking tools. If the device matches what's on the system, users gain access! Another method is called a brute force attack, which requires a combination of different characters. It executes through utilizing a hacking program! Risk is the probability of occurrence that vulnerabilities or threats will exist. A professional security role is to minimize risk to an acceptable level, a function of risk management. Learning Point: Threat X Vulnerabilities = Risks.

HUMAN INTERACTION AND CYBERSECURITY

Historically, culture and technology have evolved into single entities and created environments where humans, culture, and technology interact. Humans are the end users that utilize technology. Culture identifies the social behavior and norms found in human groups and societies. These groups instill practices, influence ideas, hold unique verbal languages or perceptions, and promote management strategies to navigate technology.

Technology encompasses technical resources to perform professional or personal tasks—such as projects, online banking, educational, or entertainment activities. Through cultural practices and organizational standards, humans may interact differently and use different technical approaches. For instance, Company A may operate a cybersecurity culture as the premier practice—while Company B may work cybersecurity as a program, which demonstrates the culture approach, decisions, and work-related tasks perform differently.

While working on various federal and DoD projects, I noticed that IT and non-IT personnel would disclaim cybersecurity. To further complicate the issue, the integration and practices were defined as a dark society. Was this the culture of choice? Often, we would have security awareness training, but to embrace security as a culture was of no concern. Could this be a result of compliance serving more importance than risk?

Corporations have historically separated security as another entity. When daily challenges and issues surface, many professionals state, "Call the security folks—it's not my problem." I never bought into this concept but believed a culture shift was required.

As a team, IT personnel work within different skill-related areas and share the same vision: *reduce risks and protect the system.* In the US Navy, this was the culture required to keep "the ship afloat." As I transitioned from the military, the same approach applied across many technical teams. We were successful at embracing a technological culture that served to protect data assets and information. Many projects did not adopt the concept, and ultimately, they failed.

While working as a security analyst, I spent hours analyzing reports and creating defensive measures for various systems and applications. As always, I embraced the cybersecurity blueprint for success—*Think Like a Hacker*—and as I began to obtain cybersecurity certifications, the same concept applied across the Certified Information System Security Certification (CISSP) examination that lasted for six hours, totaling two hundred fifty questions. This *blueprint for success* transitioned into *The Cybersecurity Mindset* and provided a career path that advanced beyond my expectations.

NOW, WHY SUCH A BOOK?

As a cybersecurity professional, I have first-hand experience concerning cybersecurity disconnections, challenges, and its blueprint. This book provides a common-sense approach toward the thinking process, mental involvement, and strategies to embrace cybersecurity. Whether your career path is directly or indirectly involved with technology, the Cybersecurity Mindset aligns with typical engagements you experience. We are all involved somewhere, someway, or somehow.

A reader will find various terms and examples of real-world explanations built upon previous knowledge and information shared. As a reader, one will master a structured approach to understanding the cybersecurity mentality and think "defensively" within the digital culture. Also, one will engage in familiar terms, processes, and experiences that highlight relevant situations where society and technology users are cyber-connected.

This book structure and information helps to articulate risks and technology as a learning vehicle versus distant details. The book title, *The Cybersecurity Mindset: A Virtual and Transformational Thinking Mode*, outlines a three-layer concept throughout the chapters. Each chapter strategizes and outlines the "cybersecurity thinking" mode. In essence, it emulates proper security practices.

If you are a professional, student, or intrigued by the word "cybersecurity," *The Cybersecurity Mindset: A Virtual and Transformational Thinking Mode* will enhance your overall knowledge-base and promote cyber awareness. The layout builds a virtual pathway to the Cybersecurity Mindset and best practices. Some may regard the methodologies as human behavior and a Cybersecurity 101 course, which is true. To fully understand people's Cybersecurity Mindset requires in-depth thinking and a technology engagement. Let's began the journey and dissect *The Cybersecurity Mindset: A Virtual and Transformational Thinking Mode.*

• VIRTUALIZED PATH •

- Inclusive Culture
- Situational Awareness
- Risk-Based Thinking

INCLUSIVE CULTURE

The most challenging aspect of technology is cultural development—as it provides the opportunity to shape security teams, staff, and non-technical professionals. The term defines how technology professionals bond and exhibit similar characteristics through their working relationships, cybersecurity engagement, and cohesion. The process can take some time and requires buy-in from managers, supervisors, and subject-matter experts (SMEs). Despite the challenges, the IT industry has developed some of the brightest talent and problem solvers. In common, everyone analyzes information and speaks a particular language. It's somewhat a programming code that grows over time and becomes intact. Before individuals communicate a specific term or security-related information, the recipient already knows what's being said and starts their engagement or response stage. It's not a negative effect to be programmed unless you are a non-cultural involver. These are merely functional teams that have no interest in cybersecurity—such as non-IT professionals. Some non-IT professionals engage and must involve themselves with cybersecurity. As history provides the best facts and evidence, non-IT professionals grasp technology as time progresses and become a cultural partner. By default, non-IT professionals are married into a technology culture driven through working relationships and curiosity. Despite which avenue constructs the cultural connection, the end-state builds a security culture and mindset that unilaterally operates.

A culture is a set of shared attitudes, values, goals, and practices that characterizes an institution or organization. Family history, college institutions, religion, and geographical backgrounds contribute to its developmental process. How a culture responds to situations and engagements represents their thinking and mental state—as they organize their social or professional lifestyles the same. Each culture can be easily identified since people display the same attributes, language, food, music, or communication style. A further definition implies that culture promotes learned behavior patterns. By default, each member behaves and promotes social norms. Once a cyber-dude, always a cyber-dude!

The technology industry constitutes a large and very complex culture. There are penetration testers, administrators, developers, or client-service professionals. Each segment shares commonality in IT—help resolve problems and advance business operations. In our personal lives, we have been culturally shaped, and within IT, the same occurs. Through a repetitive connection, IT personnel become culturally intact and harness an inclusive culture.

The term "inclusive" defines all the attributes and security requirements that encompass a particular culture. The overarching strategy describes how the corporate security personnel and program should operate through a comprehensive security image. For instance, when a security analyst starts a job, they are new to the IT environment. The onboarding process and initial team meeting profile the culture. After a defined period, the security analyst can "fit right in." Here is where the transition occurs, and they learn the IT practices, roles, responsibilities, and vision principles, which are all-inclusive to the culture. Later, the security analyst transitions to using language, terms, or security-related discussions that are culture-specific. Each is a result of learned behavioral patterns and work-related practices. These norms later become a security analyst's survival tactic—as they must fit the cultural image!

CHAPTER ONE

IMAGES OF AN INCLUSIVE CULTURE?

Technology provides distinctive elements and processes that affect our security engagements, task objectives, and team interactions. As a security steward navigates their career path, they encounter different people, methods, and techniques to sustain security. The steward may perform various tasks that require the same or modified policies as they develop many skillsets and transfer between employers, different approaches, and thinking models. The outcome provides many ideas, policies, and working relationships that describe the organizational cybersecurity profile, leading to many cultural ideas and approaches. At first, it may become confusing, but after years of experience, they become culturally prone. Having a placement in many cultures can sometimes be beneficial. The knowledge gained can sharpen technology skillsets, develop the best career path, and provide growth and value as an employee, employer, or manager, and this is where the image circulates.

Every enterprise has goals, policies, and regulations that describe its security operations and plans. The images are just that, a descriptive statement or required practice that represents its security objectives.

The standard definition for an image is a visual representation or photo of something. In the context of technology, the photos are profile statements and operational procedures that position a company to gain security success. Throughout the business lifecycle, the images are related to its core practices and operating procedures. In the cybersecurity arena, the photos serve as standards and best practices. For instance, a risk management program may require every manager to follow organizational policies for submitting a detailed report— and this serves as the business's "image" or operational profile. If there are deviations or individual reporting standards, the reporting system would be useless. Alternatively, individual reporting becomes counterproductive and misrepresents the policies and standards. So do not destroy the image—it represents a direction and standard. As once stated, standards are developed for a reason!

The cybersecurity market or cyberspace is a complicated environment that depends on experienced professionals. Their role is to protect the "bubble" and outsmart the "bad people" or "hackers." The "bubble" is where internal and external information gains or becomes restricted from accessing the technology environment. Typical terms such as network, offensive, defensive, or boundary describe the entry points. Since the bubble depends on various teams, policies, regulations, and group-based thinking, success or failure relies upon human interaction and a practical buy-in philosophy. Human involvement consists of understanding the business blueprint for success and protecting the brand. Management teams and their philosophy control the buy-in structure and dictate whether its cybersecurity culture succeeds or fails. This is somewhat a challenging task but serves as a core element to building security. Without an influential culture, the Cybersecurity Mindset is weak and useless. Alternatively, a practical perspective creates a very sophisticated image that exists across every security boundary. These images describe the content, system profile, or corporate direction. In most cases, the leadership falls short of aligning its security culture and brand, which stems from security regarded as a separate workstream and solely involved within the IT community.

The objectives for building an active cybersecurity culture require having a value proposition statement that describes the culture and its security benefits. Technical teams often isolate their position from the corporate structure, affecting the brand and business goals. The entire direction for security becomes misaligned and weakens the defense posture and image, and having a value-related approach can resolve the issue. A value proposition symbolizes where, when, and how human decisions and the corporate brand cross benefit and builds security. Many companies utilize value propositions as a selling pitch to customers; alternatively, it's an internal practice. When team members view the outcome, they can better align their positions, defend against attacks, and reduce risks. Organizations need this type of image—as it helps foster a cultural brand that remains active, responsive, and security-focused. Of the three, incorporating an environment that is security-driven is top priority. It defines security as a business driver.

In the context of many programs and objectives, business goals are a far-end thought due to internal team values or program deficiencies. Many IT security programs are unorganized and promote weak value streams. These streams are defined as excluded cultural practices, norms, processes, objectives, or tasks because they do not operate within the cultural brand. Some examples are delivering dysfunctional privacy programs and failing value-related services when the corporate culture works through a quality approach. In the inclusive culture, a quality approach alleviates program dysfunctionalities and operates an effective value-related service. Before releasing or engaging cybersecurity services, the corporate quality approach is emphasized and becomes inclusive. This is where security becomes aligned and enhances the cultural image, and the outcome provides structure. If an organization works in the opposite direction, risks are developed, and management teams create challenging situations.

Although management may find abnormal and challenging situations, the focus is on security pillars: Humans, Information Sharing, and Technology. Humans are frequent users of technology and individuals that use email communication. Information sharing is a technology concept where information or data exchanges. Sharing can occur via email, smartphones, computers, or a web-based resource. Technology is the vehicle that enables humans to use computers. It also allows data communication via online resources and provides a protection mechanism to ensure data and information protection exists. These pillars rely upon human decisions and their behavior.

Management strategies, team development models, and work-related processes can influence behavior. How these resources collaborate determines the culture's cybersecurity practices—such as using risk reduction and mental models to align thought processes, positional involvement, and practical tasks. Every involved team member approaches and values the corporate brand, but not to the same extent, partly influenced by failing cultural practices and experience. If an employee is accustomed to practicing and utilizing behavioral techniques that seem correct, they will extend the practice between employers or role assignments. It makes no difference whether they are a security engineer, cloud controls analyst, or risk manager; the behavioral practices can induce additional risks. A great example that demonstrates the concept is a vulnerability remediation task. A security steward may interpret mobile vulnerabilities differently from their coworkers, which could stem from task involvement practices that excluded the security steward. A shift within the task responsibilities would sharpen the security steward's mental approach and enable an environment where every mobile vulnerability received equal importance. Reducing human risks can be more powerful and beneficial for the cultural image.

Many definitions describe human risks. As discussed before, they are formed from the three security pillars. A traditional description always examines technology and configuration settings or broken defenses. Within the mindset model, it's described as not thinking or engaging security best practices. It's straightforward that human risks affect technology and cultural norms. Technology interaction and decisions are technically involved, while cultural norms are the designed procedures throughout every program or system, and each responsibility is to improve security. The process requires a top-down approach that understands the organization's fabric, and employee involvement, starting with the Chief Information Security Officer (CISO).

The CISO is a high-level authority that drives the corporate security vision and strategy and executes employee involvement. They spend much of their time securing the technology environment, people, resources, and assets. Their position comes with increased knowledge and understanding of the entire enterprise. Every organization has a CISO, and they usually serve and direct cybersecurity across many platforms and services. They strive through all challenges, causes, changes, or differences to exhibit a cyber-safe environment and a functional security culture, which is challenging.

CISOs encounter cultural challenges that create risks or security gaps. The responsibility and ownership surface with a straightforward answer: *a disorganized culture can cause security gaps.* Sometimes these issues are digested and modeled toward continuous improvement, and then there are situations where gaps exist. When this occurs, it demonstrates a corporate culture requiring change, attention, due-care, or a disciplined approach to driving safeguards and defenses. Through a value proposition model, practical business approach, or new management strategies, breakdowns could be resolved. These solutions balance CISO goals and employee satisfaction. When formed correctly, a functioning setting would exist where employees operated toward specific security goals, shared a concrete philosophy, and are mentally aligned. Here is where a CISO can find the best return on investment. Alternatively, a dysfunctional environment works in the opposite direction and may require the development of a cultural mindset.

The cultural mindset requires a technology approach that onboards every safeguard and defense tactic. When developing the cultural perspective, one of the challenges is integrating ideologies that seem exclusive and risk prone. A further example is doing it "our way" or "what we see as the best practice." In some cases, it's appropriate, but does it align with cultural goals and policies? As a security steward, it's a duty to question work-related methods that exclude critical security standards. All too often, the cultural identity fails this task—as politics and bureaucracy carry more importance. It further creates misalignments for various safeguards, defense tactics, and protection schemes and allows hackers to access the system quickly. Having this image is not something any organization should possess or tolerate. What should be accepted is a culture where leadership exhibits a commitment to inclusive practices. These embedded strategies are internally designed and addressed as norms or "functional objectives." Typical terms such as "how we operate" or "our objective is to…" are an initial approach to the best security practices and security image.

As a CISO, you have mastered a strategy to remediate cultural challenges, and you feel the threat landscape is safer. Well, there are additional cultural challenges to address. Considering there exists a diversity of teams that work independently and dependently, a broader challenge does exist. Within any IT division, business unit, or group, organizational interdependencies exist. You could have database administrators, help desk, network management, or technical teams that coordinate operational goals. Although these different teams exist, they operate inclusively at mitigating security threats and vulnerabilities. Inclusive is defined as the internal scope, landscape, and processes that encompass the corporate defensive tactics and responsibility. To survive, each security program structure, processes, and workflows operate through their designed interdependencies. By default, organizations structure their interdependencies due to projects, operations, and business goals. Within its structure, the cultural identities are established, and as the team develops, they become interdependent. There is no magical formula or preexisting plans to develop interdependencies. By default, the groups operate through interdependencies, which forms a term called "interdependence of operations."

The interdependence of operations exists as a management strategy that drives teams to resolve security deficiencies via cooperative relationships. Each team knows their security responsibility depends on individual roles; mutually, both operate toward a specific technology goal. In 1967, sociologist James D. Thompson discussed the relationship in a book called *Organization to Action*. He wrote that many groups and teams work along independent lines but interact to address dependencies. These dependencies can be goal-driven activities or solutions. Many teams operate and share information through separate functions, dependent work tasks, coordination, or input and output methods due to interconnected business processes and working relationships.

The interdependence of operations can transition into the technology space and its service model. Various interconnected systems and workflows rely upon information input and output (I/O) and separate functions to execute technology-related tasks. As projects expand and different security objectives are designed, networking, operations, cloud, service lines, and penetration teams must collaborate and maintain their responsibilities. Each operates and delivers individual services that support a high-level task. Inclusively, the different groups model the cultural image, such as practicing security and business as one entity. Various communication practices are either independent and dependent, and project objectives are described for each team. A great example is an inclusive culture that engages incident response tasks and goals.

Incident response is a coordinated effort where many teams and tasks coexist. Each IR phase may differ between organizations, but the organization's overall goal and team alignment mirror. The requirements for networking, database, helpdesk, and coordinator duties are consistent. For instance, when significant attacks occur, the data analysis process coordinates database administrators and separate teams. The networking team relies upon information that details the attack's source address—so the helpdesk personnel must assist with system identification. The database administrator needs to know whether a database attack occurred, and the SOC team has to provide the DBA the attack type, source, and destination. Last, the incident response coordinator does their part: coordinate the analysis and data collection from all teams. The attack's remediation is the goal while each team independently performs incident response tasks—a perfect example of the cybersecurity image.

Now, what type of image should the cybersecurity industry demonstrate? In terms of mental alignment, the workforce should exhibit traits where conformality, buy-in, adaptation, and security thinking cooperate. Of all the alignment practices, a buy-in carries the most weight. Here is where unwritten agreements and inclusive thinking surface. The corporate culture does not write contracts or challenge independent thinking because its cultural image alleviates the concern. Employees embrace the brand and model security: *"We Can Think Before Hackers."* Another model is corporate responsibility, which is where organizations implement IT defensive measures and onboard responsible actions that support stakeholders and strategic goals.

The strategic goals depend on having an influential leadership culture. This model operates well in the technology space and creates ideologies, processes, and tasks that build the Cybersecurity Mindset. When the leadership exhibits these core principles, the outcome aligns with system defense standards. The alignment is where responsibility, authority, and accountability exercise and supports risk closure. Responsibility is seeing through the entire task, and authority is simple: a leader must assume charge for the security culture. Last, accountability states, the leader must have an answer. When all three are practiced, leadership can remediate the cybersecurity culture's challenges and build a win-win relationship. Also, their presence can influence a "buy-in" structure and drive responsible security, active involvement, and tighter relationships for business lines that manage security services.

Leadership is one component that builds the cultural image. Beyond its contribution, policy enforcement is required. Policy enforcement can detect violations and outline actions. Its use is administratively driven, but its outcome can change and enforce operational and technical requirements. In 2018, the ISACA Cybersecurity Culture report played a pivotal role and expanded information for the policy enforcement role. The report outlined that 58 percent of corporations do not have a cybersecurity management plan or policy. These plans or policies integrate different cultures and management strategies to combat cyber-attacks and risks. The lapse in these policies affected the cultural mindset. Employees are working with no vision and structure, which damages the cultural existence or development. Based on similar experience and knowing how solutions operate, having a good leadership team that "buys into" the policies is more beneficial. Alternatively, the opposite will create identity issues and separate the culture.

The leadership role drives the functional teams and buy-in for ownership, responsibilities, and regulatory policies. Any strategies outside the concept will separate security and cause safeguards to fail. This leadership strategy is a business risk considering how breaches, system intrusions, and network failures occur. Imagine a CISO excluding technical teams and business lines from decision-making exercises. Would they gain a sense of ownership and accept responsibility? In most cases, they would not consider the option.

A CISO's decision affects the risk posture. When particular risk areas and recommendations are excluded, they affect the enterprise, and critical systems become risk-prone—this is why culture involvement and cybersecurity are one entity. Eventually, the CISO decision-making would result in security gaps. Although the scenario sounds simple, its impact is very harmful. Just one wrong decision can navigate more risks and "feed the hackers' appetite." There are valuable solutions and management methodologies that would support the scenario, but remediating human threats is more effective, which is achieved by integrating a "human firewall."

For many years, the term human firewall has existed within the technology industry. It's used throughout many services and engagement strategies where people think protection and act defensively. The methodology is more mental than physical—as humans utilize mental warfare over physical warfare to protect enterprises. A human firewall is not a complicated methodology; instead, it's an innovative concept that defines human interaction, access control, and system defense strategies. Another terminology for human firewalls is a defensive mindset. When roles and responsibilities buy into the human firewall principle, they foresee themselves as the security culture. They envision an environment where accountability and authority resonate through working habits, goal-driven activities, and functional teams.

While working on federal government contracts, having a functioning culture entailed that government and contracting personnel supported a single vision—this derived through a win-win team concept that summarized the IT culture. Each analyst demonstrated the characteristics of a human firewall. You would often sense a "restrictive" attitude amongst the team. The external teams would have to authenticate information, and data sharing followed secrecy and confidentiality. Imagine calling an IT personnel to change a password. As a human firewall, they would verify account information before proceeding. However, the process was pressing and tested people's patience.

The government utilizes the human firewall theory more often than the private sector. The difference is drawn from data protection and classified information sharing. In every federal agency, there exists a security clearance requirement. Each contractor, federal employee, or vendor must have prior authorization before accessing sensitive networks or IT systems. The system administrators, system managers, and security officers position themselves as human firewalls when access is requested, and they question or require information concerning needed access. Every contractor has a story about gaining clearance and will state; it's a grueling situation! Once the clearance is granted, the next hurdle affects information sharing. Although someone has been granted system access, information sharing is very restrictive. Try asking or requesting information, and you will commonly hear, "What team are you assigned?" or, "What's your clearance level?" These are perfect examples of a culture that perfects the human firewall theory. It works best when employees, management, and technical teams function as a singular vision.

A singular vision is a cultural identity that represents the same goal or end-state. Each technology engagement follows a described concept for safeguarding, protecting, or mitigating risks. Each vision defines the organization's operational procedures, such as defending critical applications or controlling access. One significant cultural identity that exists is the reaction to cyber-attacks. When an incident occurs, other IT personnel collaborate and strategize safeguards, countermeasures, or kill-chain strategies. The goals and security mindset are enacted with minimum direction. Although external teams operate independently, they share a cultural identity, which is to defend the enterprise.

Without a cultural identity, how would an organization respond to cyber-attacks? First, the risk of furthering system damage would exist, and the response team operates separately and causes many distractions. Imagine a risk issue such as a worm or trojan spreading from 10 to 50,000 users due to the incident response team's carelessness. Afterward, the "shift-of-blame" begins to surface, and finger-pointing arises. The issue can be readily resolved via a centralized management strategy. The centralized approach builds proactive responders that understand the cultural vision. The concepts provide all IT personnel and business units with an identical and mental system to think defensively. Alternatively, the decentralized method allows the teams to function independently, directly affecting the corporate brand and reputation.

Corporate reputation is the opinion customers and end-users have about the security services offered. Organizations may hold a high reputation when their cyber hygiene is healthy, responsive, and valuable. When reputational damage occurs, it poses many ramifications and impacts on the corporate brand and customer confidence. In November of 2018, the UK telecom industry released a report on cyber-attacks that addressed reputational damage. The report demonstrated that the time required to respond and mitigate attacks was very long. High-level DNS attacks and mitigation timelines were slow and induced additional risks. The attacks increased to 42 percent over twelve months due to the slow response and remediation process. If there were a 100 percent committed effort, the delayed response, mitigation of vulnerabilities, and infrastructure hardening would have had success and remediated any concern for reputational damage.

When the reputational damage is remediated or prevented from occurring, the business can operate a functional culture—different security teams working under one division or goal, which means that network, risk management, vulnerability management, and cloud security cultures operate as one. A great example of a functional culture is a security operational center (SOC). A SOC is a facility that houses an information security team that continuously monitors and analyzes an organization's security posture. Each personnel demonstrate language relating to hacking, incident handling, or computer network defense strategies, despite being a network security or risk manager. The same and use of the applications exist. Each staff is working toward sustaining operations, keeping a system safe, making regular reports, and resolving issues while sustaining an inclusive culture.

The inclusive culture can remediate cybersecurity challenges through mentally aligning teams, resources, and engagements.

When successfully practiced, each organization can drive a secure environment that's goal driven. The SOC environment is an example that mirrors the methodology and produces shared experiences and cohesion. The entire culture shares a singular vision and models the corporate security image—protect the enterprise. When risks surface, they independently and depending perform. Imagine an environment where a malware attack occurs, and as a collective unit, the business remediates the attack and sustains operability. The outcome can reduce monetary loss and maintain public confidence—driven from understanding the enterprise standards and acting upon them when needed. Also, the results affect public confidence and drive cultural inclusion. It further translates how the customer base interacts with the business and whether both share the same vision. If they feel the company has or will fail its obligations, the culture will collapse. To counter the practice, security organizations can address growth, maturity, and development, which symbolizes a growth mindset.

CHAPTER TWO

GROWTH MINDSET CULTURE

I n 2012, the CISSP certification expanded its worth across many organizations. Many of the career-breaded professionals attended training courses to learn CISSP-relevant course information. I remember working on several IT projects and learning about the CISSP certification and its recognition. As an eager professional, I dedicated time and energy to the required training, and it was not the most straightforward certification. Many of the test questions were foreign and seemed vague. After pursuing the examination and achieving a passing score, I transitioned into a challenging job market. The certification has achieved more value and considerably grown due to technology maturation and varying skillsets requirements.

During the transitional period, the IT security culture expanded. Various technologies surfaced, and security experts found themselves studying and attending conferences so their skillsets and training could remain current. Although the culture embodied technology, there was a gap in understanding which professionals represented IT security. As the industry continues to develop, the same question surfaces. How should workforce development be structured so a cyber-centric environment exists? Should the cybersecurity culture change, and at what pace? Considering technology, business needs, and requirements change daily, a cybersecurity growth mentality is an answer.

When a growth mentality exists, the cybersecurity environment can develop, reshape, reform, and transition. One example is the emergence of cloud technologies and how it has strategized the cybersecurity maturation process. Millions of online periodicals and business solutions grew from the cloud initiatives and their technology innovations. The entire web is saturated with cloud-based data and information, where the cloud has become the "talk of the town." The technology innovations have shifted IT environments to scale their operations and move remote responsibility to the cloud service providers (CSP). In the earlier years, monitoring and maintenance responsibility was far higher when using local service; now, the CSPs hold the "sword." A unique language is also used for virtualization, subscription services, and service-level agreements (SLAs). Outside of communication, there is a dire need for new certifications and training. Can you see the growth pattern? The cultural identity and growth mindset shifted from average IT to cloud-based solutions.

The term "growth mindset" is not a new technology exposure. The methodology has existed for many years but is rarely discussed within the technology space. Its initial interest transpired from a world-renowned Stanford University psychologist named Dr. Carol Dweck (Mindsetonline). She published a book titled *Mindset: The New Psychology of Success* to enhance teachers' and students' education. The methodology emphasized that people who practice a growth mindset pursue dedication and productivity and build better relationships for business, schools, and sports. Dr. Dweck furthered the definition by saying, "Intelligence improves through study and practice."

In *Mindset: The New Psychology of Success*, Dr. Dweck introduced a counter-theory called fixed mindset, which reiterated that talent or intelligence was all-inclusive and the pathway to success. It sounds more like an organization has become comfortable and wishes to pursue success through skill and knowledge. When comparing both mindset conditions, the growth mindset possesses intelligence and talents that maturate or developed over time. A fixed mindset is stagnant, and there is no desire to promote talent—as expertise outweighs skillset development. Thanks to Dr. Dweck, we have a formalized methodology, the growth mindset, that develops a skilled culture.

The transition from Dr. Dweck's methodology and the growth mindset models the cybersecurity culture. Various organizations perform security assessments, infrastructure hardening, and system sustainment. These functional areas seem to produce metrics and compliance benchmarks that present achievement and effort. Although organizations may score a high percentage, the need for maturity exists. The requirement is a security alignment. As the industry evolves, so should organizations' security practices or processes.

During the IT career development process, I have seen many well-to-do professionals concentrate on intelligence and undoubtedly become sand-boxed into their career paths. If you question these individuals about achieving growth, you will probably hear, "Look at my role now." Relating to Dr. Dweck's theory, intelligence can be a significant distractor and fail growth and maturity—partly due to management teams and organizations feeling they have achieved considerable security progress. When examining many organizations, we can always discover gaps, weaknesses, and areas requiring improvement, which negates the management theory that they are secure. In 2014, I faced a similar situation while attending a meeting. Upon taking a break, a businessman approached me, and we started to discuss our company's service offerings. Upon learning about each other's business profiles, he posed a fascinating question. He stated, "If we are not under attack and our operations are going as planned, would it be safe to assume we are secure?" Being a cybersecurity professional, I knew this was a broken rule. I kindly responded and said, "You are not one hundred percent secure." The gentleman looked at me and also wasted his coffee. I stated, "When your car has a malfunctioning part, you can drive several miles before it fails," and the same concept mirrors IT security. Sometimes viruses or malware are not discovered; when they become active, the outcome can cause severe disruption and business loss. This example is not designed to illustrate the gentleman was illiterate toward IT security, but it does stipulate he needed to develop his growth mentality.

Cybersecurity development is a broad and complex term. There is a friendly relationship to several working models, compliance initiatives, and requirements, such as the Cybersecurity Maturity Model Certification (CMMC) framework, which is a capability maturity status for contractors when servicing the government. Each contracting company must assess and reevaluate their cybersecurity growth and development process or risk not having government contracts. Similar to the CMMC, organizational cultures that practice development often research and find innovative ideas to improve. As the organization develops its capabilities, growth must fulfill technology. Imagine an organization that builds a framework where practices are stagnant, and their focus was on achievement vs. development. Considering new attacks are surfacing daily, those same organizations would become "a hacker's appetite."

Organizations must develop and grow ideas to defend systems. When attacks happen, best practices and lessons learned provide ideas and areas for improvement. A counter discussion is when attacks do not occur. At this point, most organizations are probably at a fixed mindset and thinking: our systems are safe. A growth mindset is an opportunity to develop and achieve. If organizations remain fixed, new attacks and hackers are winning the race. The intelligence in understanding the latest attacks is behind industry knowledge, and hackers are having fun!

When thinking about Dr. Dweck's research and conclusion, the history of working on various IT projects and the growth mentality are familiar. Upon starting the career path into IT, I experienced many disconnects and focused on network security. As attacks surfaced and applications became more vulnerable, organizations extended talent to build resilient systems. Also, application code scanning and different tools emerged. The initial process focused on network perimeter devices such as firewalls, routers, switches, and hubs. Today, the industry has transitioned and developed tools, techniques, and technologies to address vulnerabilities outside of the network and perimeter defense.

Cultural growth derives from risks, attacks, and public confidence, but financial costs also contribute. As system protection and sustainability costs stem upward, organizations must develop cultures that improve technology. While working on a MacAfee project in 2009, I experienced new development for system security. The Department of Defense onboarded McAfee ePolicy Orchestrator (McAfee ePO) or Host-Based System Security (HBSS) as its defacto CND tool. It incorporated modules to identify vulnerable actions for hosts (endpoints). The DoD deployed HBSS to counter data breaches and system hacks. The emergence of HBSS transitioned the DoD security and intel culture to think "cyber defense." The traditional culture focused on safety, but were they safe? Cyber defense involved examining threats and preventing attacks, and strengthening cybersecurity—while safety focused on protection standards.

The Department of Defense (DoD) has made similar advancements for growth and protection standards. In 2019, it published a comprehensive digital modernization plan to advance and mature its cybersecurity landscape. The plan emphasized consolidated infrastructure, streamline processes, and workforce strengthening requirements. During the phase, the DoD onboarded cloud technologies, unified systems, facilitated compliance programs, and transitioned to joint enterprise architectures. Some of the techniques were new, and from a technical perspective, the growth mindset was active. Relying on older technologies, outdated processes, or fixed state mentalities was never a concern. As the modernization plan materialized, the DoD updated policies and standards to complement its enterprise architect and culture.

As organizations consider growth, policies and standards must align. The objective of the alignment is to reduce enterprise risks and integrate protective schemes. Legacy policies and rules may not provide the resource to achieve the goal. Think about security programs that rely on legacy standards. The plans would induce more risks and create process gaps, and the risk expansion would surface and become a reality.

A process gap addresses a lapse in best practices defined as a missing security feature, but the organization can still operate. A perfect example is how SOC teams report incidents. If an organization's malware attacks change over time and new malware exists, its incident response procedures will require new counterattacks plans. The procedures for using legacy standards, processes, and policies may not provide the resource to understand further attacks, and the SOC team may resort to researching during the attack. Incident response is about "how to become smarter than hackers!"—so reaching is good, but better when performed ahead of or before attacks occurring.

In 2018, Gartner Research published a periodical titled *The Future of Incident Response: Changes, Challenges, and How to Prepare.* The article discussed how the changing landscape, threats, and incidence response must develop to counter emerging attacks. Gartner research and growth mindset are one entity. When performing risk assessments concerning security gaps, a security assessor must address new and emerging threats. These threats force our current incident response procedures to change and include new systems. All too often, there exist instances where further or advanced information is gapped or omitted from security taskings. Although the task may become challenging, the result is risk positioned—as the inability to defend the system could result in process failures. For instance, cryptojacking is an emerging threat where an attacker hijacks a computer to mine cryptocurrency. CryptoJack uses a victim's computer as a computational device to update cryptocurrency blockchains. New tokens creation occurs during the attack, and deposits occur using the tokens. Imagine having a fixed mindset in this case: we are not susceptible to crypto-jacking. The growing cryptojacking issues would be overlooked, and then, *"Here Come the Hackers."*

Gartner's research directly relates to CISO shifting its management strategy. The shift requires CISOs to stay abreast of industry trends and attack profiles. In relationship to the inclusive culture, it's essential the CISO exhibits and promotes a buy-in process. Most employees fear change and growth-related processes. This is a direct belief that new policies, guidelines, and legacy processes pose severe challenges, and performing outside the normal comfort zone is painful. As the CISO sets the vision for onboarding a growth culture, the belief will eventually change, and so will their strategy.

As the technology industry and governance teams partner to resolve security issues, it's essential that the growth mentality and leadership strategies mirror. Management must develop a comprehensive vision that supports both approaches. The security features, services, emerging attacks, business requirements, and technology innovations are the core elements. Management strategy must balance maturity against the enterprise strategic plan and set a clear vision.

Every manager has faced a growth period. How the manager approaches growth depends on risks discovered and technology concerns. When leveraging risk into the process, a manager requires full vs. partial visibility into the enterprise program, which means that management has proactively monitored security beneath the organization's surface. You could describe the surface as just reporting results. There may be a lapse or reporting inaccuracies through various data quality management schemes, so the manager must inspect and verify the data received. Another component derived from reporting is cyber safety—this is where the fixed mentality exists, and it would be a failure to say, "Our environment is cyber safe." The dependence on previous knowledge and intelligence does not constitute a cyber-safe environment—as they both could change and induce harm. Staying current and relying on the mental mindset overrides old intelligence.

When setting a vision and navigating growth, managers should integrate emerging technologies, which depend on an organization's architect, infrastructure, and technology. Since the inception of cloud technologies, there has been a rapid increase in technology growth—it has bypassed the fixed mindset mentality and channeled technology to accept cloud technologies. Cloud computing provides subscription services and computer technologies using network access in a shared environment. The original architect used for business systems housed servers, firewalls, computers, and various software packages. Through cloud technologies, these same services exist, except the cloud service provider (CSP) is the owner. Its migration provides solutions, computer services, and technology usage through shared, scalable, and on-demand services, which means an organization can order subscription services and scale them upward or downward. The traditional local area network (LAN) required businesses to purchase equipment, software, and additional services. Through cloud technologies and integration, companies purchase LAN services from the CSP, which lowers maintenance and operational costs.

A great example of cloud usage is during the Christmas holidays. Online retailers purchase cloud infrastructure services to operate their applications. The purpose of having these cloud services is to support the rapid sales and flux of data transactions. The CSP infrastructure houses the online retailer applications, which shifts the maintenance and operational cost to the CSP. After the Christmas period, online retailers downscale the use of cloud services.

Cloud technology is also a component of management's vision—as it helps reduce risk and induced costs. The government market incorporated the cloud under several programs and strategic ideas. The intent was to bolster technology, save money, and align services. The alignment resulted from a "cloud-first policy" that required federal agencies to consolidate IT services and operations, and because the government spent over $600 billion in information technology from 2000 to 2010 (Plan, 2010), the cloud-first policy was a national initiative to reform the Federal Information Technology program from 21,000 datacenters to 800 (Plan, 2010).

The expansion of the cloud-first policy extended to the Veterans Administration (VA). The adoption supported the VA personnel, contractors, partners, and veterans to access information and services. The traditional architect housed a multitude of data centers, applications, and services. Through understanding technology innovation, the VA integrated cloud services in 2012 to align their services with the federal needs: develop a robust, reliable, and safe infrastructure that models growth. The initial cloud-based platform integrated Microsoft Office 365 for over 600,000 users.

The cloud-first policy also enabled alignment and positioned technology, management, and federal agencies to overturn a fixed-mindset mentality. Under the fixed-mindset attitude, growth was excluded and undervalued. There existed a stagnant and redundant pattern of managing 21,000 federal datacenters. Prior engagement consisted of adding infrastructure and services to resolve security issues, which defeated developing a robust and safe infrastructure model. A significant point was proved: the fixed mindset mentality influenced the cloud-first policy for the federal IT market. If the fixed mindset remained, the federal government would have to sustain the operation of 21,000 data centers; and the entire federal landscape, threat posture, and risk exposure would have expanded. The cost to maintain the datacenters would increase, and *"Here Come the Hackers."* The cloud-first policy aligned to the Department of Homeland Security's (DHS) *Twenty-Five-Point Implementation Plan to Reform Federal IT Management.* The plan stated that there existed little productivity through the current IT architect in 2010, which prompted changes in federal IT spending and design to "Do Better Business."

Business improvement is a procedure that enables program development. The process requires a dedicated team of professionals who research, design, and implement activities that enhance or reduce organizational risks. The outcome promotes business process improvement (BPI) and an environment that mirrors maturity. The functional areas that enable BPI to operate are performance, productivity, and quality.

When referencing previous experience and exposure to BPI, the relationships and technology innovation are standards that grew from BPI. Many organizations perform security assessments and never consider the opportunity for success. You can visualize "opportunity for success" as a point to improve outcomes. When reviewing the process for security assessments, the results represent a failing point. Still, when referencing BPI, it's a significant area that allows organizations to define weaknesses and make improvements. There was an old saying: "If you have no idea where you are at—how can you move ahead?" On the contrary, when risk state performance and quality are unknown, can organizations successfully profile their goals? In terms of quality, the structured plans may have significant gaps and affect additional BPI considerations.

A growth mindset has various elements that support maturity and development. The creation of the NIST Cyber Security Framework has been an international enabler. Countries such as Italy, Bermuda, Japan, and Israel have onboarded the framework to prevent attacks and manage technical risks. There has never been an international standard created with clear and concise information for the public and private technological sectors. Frameworks such as ISO 2700 leveraged great attention, but the *NIST Cyber Security Framework—Improving Critical Infrastructure Cybersecurity* gained better attraction. Since its inception in 2014, the framework matured and onboarded a Cybersecurity Enhancement Act and received a new revision called NIST Cybersecurity Framework V1.1.

What if the framework was nonexistent? Over the past five years, the number of hacking incidents, data breaches, and issues has grown; and with the framework's existence, growth was successful, and risks reduced. With a nonexistent NIST framework, the US safeguard alignments would be unbalanced, and risks would outperform countermeasures. The discussion does not define that the NIST framework was a complete answer, but it does demonstrate the necessity and how important frameworks and a growth mindset can influence maturity. Numerous frameworks provide the same results, and as they evolve, the growth mindset must stay relevant.

In the technology arena, growth happens every day, and the goals must outperform challenges. You can relate challenges as risks, hacking, data breaches, program deficiencies, or performance factors. How challenges are overcome dictates maturity, and this is where we measure growth and describe goals. Every day the technology market changes, and then the end-users or management teams are affected. It's the IT security professionals and the cybersecurity ecosystems' responsibility to develop and align best practices. To counter the change process is a nuisance and distractor. Stay vigilant. Let's create a cultural mindset that embraces growth and supports technology change due to innovation, incidents, and advanced protection schemes.

CHAPTER THREE

EMBRACING ORGANIZATION CHANGES

Securing enterprises is far more challenging due to different onboarding and programs required to counter attacks and data breaches. The onboarding incorporates new ideas and requirements where organizations restructure their defensive approach. These approaches lead to work-related tasks and involvement that change the culture. The discovery of new technologies such as blockchain, automation, or the Internet of Things (IoT) may come into existence and force varying security transformations. Last, the technology approach and its integration can create risks and security challenges. These are due to new applications or security devices being added or misconfigured. Despite the outcome, securing enterprise assets is a priority.

An organization's defensive mindset gains success when security is strategized and modeled. The strategies address a "how-to" and enable the best practices while the models mimic the end design or outcome. The results may vary and create risks for the IT systems due to new business requirements or technological changes. The shift can be very intense and requires a coordinated effort from various resources. It could require firewall teams to update security rules using new architectural standards or risk management experts to evaluate more recent threats and vulnerabilities. The entire cycle can occur over several periods or when events or incidents occur. The transition affects the current architect, and security will function as long as an inclusive culture remains aligned.

The security lifecycle approach onboards change management but rarely addresses embracing change. It assumes that the organization and its personnel had accepted the security roadmap when it's communicated or executed. Each describes the change process as being followed and bought in by the organization. It's very simple to explain and navigate the writing process, but a responsive environment that embraces change is more beneficial and challenging. It stems from people's motivational mindset and whether the business has successfully sold the outcome. Yes, people will follow the roadmap when success is the end-state. So, early and often, newer business pitches must embrace change for the corporation and the teams, groups, or employees. This is usually a concern for many managers, and it never seems to cease. The cybersecurity culture can have its challenges as well. Issues spanning from business operation, project deliverables, newer technologies, or rules of engagements can shift a person's security perspective. Gaining acceptance is a lifecycle approach that can enhance the change process and drive teams to embrace its outcome.

In 2013, the Department of Defense transitioned its risk management program from the Defense Information Assurance Certification and Accreditation (DIACAP) to the Risk Management Framework (RMF). The changeover was implemented to standardize, align compliance, lowering costs, and enhance system protection. The initial transition was not favorable and required massive reengineering. The DoD Information Assurance (IA) community had to onboard a new risk management program that required three hundred–plus security controls and assessment practices. You could hear many concerns that more work was being created—DIACAP was much simpler—or the DoD was not a federal entity. The cultural shift not only affected technology, but the buy-in structure as well. Eventually, the transformation worked, and DIACAP became history. Of course, today, many DIACAP fans think RMF is costly, a failure, and a waste.

Newer solutions require organizations to embrace technological change and innovation. The security mindset is one of the most overlooked concepts when integrating change. The fault does not imply management has failed—instead, it symbolizes that a culture's thoughts must include a comprehensive security approach and a buy-in process that "speaks" defense. There is no simpler approach and explanation since technology is ever-evolving and bringing forth new requirements. Every IT segment deals with the idea, and by far, they survive. One survival area is cloud technologies, and many business units are operating ahead by changing their technology platforms. The culture has shifted from onsite to remote protection—such as, "How do we ensure protection exists in the cloud?"

Before cloud technology surfaced, traditional networks relied upon internal infrastructure and software. The standard defense practice required in-house expertise and configuration. Since acquiring cloud technologies, the norms transitioned to subscription services, which require mutual agreements. IT professionals refocused traditional models that required internal collaboration and transitioned their approach to the cloud with innovative security thoughts: will the provider align standards that protect data and articulate data privacy? As a cyber-safe professional, these are very thought-challenging approaches that would require transparency. Imagine going through a process where newer responsibilities must be accepted. However, the results depend on another vendor or service line. Could one make sound decisions and embrace the organization's change plan? In most cases, the affected group or person would remove their responsibility since clarity or trust did not exist.

Beyond a cloud initiative, cyber-safe approaches exist in additional programs—such as cyber tool integration. Security analysts can become comfortable at using Splunk, Nessus, or ArcSight. When a culture change occurs, these tools may not provide the required analytics or visibility. The outcome stems from incidents, hacks, or significant data breaches that ultimately affect the technology culture. The business may decide to utilize proprietary or internally developed applications, which means new skillsets are required. If users are comfortable having fixed skillsets over embracing change, they may find it challenging to transition management strategies. Another failure is that the enterprise may not achieve continuous visibility, which helps gather knowledge about system assets, data, and resource protection. If there is a lapse in the visibility process, the organization could inherit risks.

An essential element for the continuous visibility program is to integrate change and cyber defense tasks. These are the components within the technology culture that define security protection and confidentiality, integrity, availability, practices, and change management as a security thought procedure. When all are achieved, a continuous visibility program has its rewards. The integration succeeds by delivering technical teams the capability and resources to identify risks within the change process. The end-state integrates risk prioritization through security involvement and removes barriers concerning risk mitigation, and teams can embrace change and think of "continuous visibility."

As a security analyst, I experienced tasks and role responsibilities that involved continuous visibility. One perfect example was when a DoD contracting company changed employees' labor categories and job tasks. The DoD wanted analysts to perform deep-dive security analysis and reporting. The initial task requirements separated the available work between monitors, reports, and responders. We spent countless hours attempting to formalize and integrate administrative reporting and tasks. Every team member expressed concern that we needed to think parallel, which required two duties:

continuous visibility and implementing change requirements. Our beliefs coincided that priorities should still protect the system assets, resources, and data. I remember several team members complaining that integrating the available work was labor-intensive and induced additional risks. Management struggled for a 100 percent buy-in, and the process was challenging. Eventually, the integration of the three teams was completed without any significant incidents. When we conducted a lesson-learned activity, the main concern was, *Did we perform as a business?*

A security culture survives when technology and business can operate in unison and outperform risks. Some organizations separate their objectives. The process allows management to define IT as just a supportive element and not a business driver. If organizations built technology as an enabler, a change could progress with fewer risks, such as embracing newer processes or security objectives. All too often, IT is labeled as a separate entity, and when security changes exist, the separation becomes a significant factor. The goals must include and view change as a needed factor vs. just an IT headache. That's right; it's labeled as a headache because it can adversely affect business operations. Specific programs such as the Information Technology Infrastructure Library (ITIL) strive to strengthen the relationship. Although it is helpful, some security programs remain dysfunctional. We could argue that it's the IT or business fault, but ultimately it's understanding and respecting why change exists.

In today's security environment, many organizations have shifted to baking security as a critical objective. The strategy requires business lines to integrate security into its thinking model and program goals. This is a significant plus to security because its development process is mutually aligned with the corporate goals and objectives. Despite falling behind, most organizations buy into cultural change and realize it is easier to accept the outcome. One solution that has supported the effort is behavioral influence. Behavioral influence is a mental approach that management and staff members must embrace and sometimes change to safeguard information assets. It's considered mental since human behavior is at attention. Technical and non-technical professionals will commonly admit that IT changes are challenging and express dislikes. The thought process is risky because systems, networks, applications, and assets require critical attention. Through social engineering practices, hackers can use a disgruntled employee as a resource or entry point. If an employee voices their concern publicly or provides valuable information where hackers can deduce sensitive information, a hacker has just "won the lottery." We live in a society where even a minor security gap can disrupt systems and networks—let's not feed a hacker's appetite.

In 2019, a survey conducted by Deep Secure found that almost half (45 percent) of office employees surveyed would sell corporate information to external organizations. The finding is alarming since organizations instill confidence and loyalty in employees. Although visibility exists toward the employee, there exist loopholes that can be penetrated—such as an employee communicating outside the corporate email system. Here is where accountability and regulations are required, which means companies must promote due diligence and implement safeguards.

The usage and enforcement of policies such as "an acceptable use policy" or "accepted rules of behavior" can strengthen and influence change. Since they are organization-specific, the procedures can enforce confidentiality, usability, integrity, and data protection practices. Aside from these policies, extensive or more vigorous behavioral methods such as a DNA framework must exist.

The understanding of corporate cybersecurity DNA is a valuable vehicle toward behavioral influence. The concept brands practices where technical and non-technical teams understand change benefits and branding. In medical science, DNA applies to human genetics and their mental and physiological makeup. Genetic makeup has genes that determine how humans think and develop physical features. The genes are inherited characteristics or traits held from parents, grandparents, or great-grandparents. For example, a tall parent will father or mother tall children; their thought process and thinking characteristics will pass down to offspring.

The medical DNA methodology is transferrable toward cybersecurity and aligns across various technology practices. When examining a cybersecurity architect, there are makeups (genes) where an organization develops methods, programs, defensive schemes, and behavioral characteristics. The development starts with connecting the security architect and transitioning the safeguards, protection schemes, and governance strategy across departments. The DNA produces traits or behavioral patterns where technical and non-technical teams think defense and know where system protection operates and value change. These security traits are communicated through new staff members and remain during transitions or rebranding. If the DNA remains active, a corporation can defend against attacks and sell change.

Active DNA is an ongoing effort that remains in an organization's fabric. The features model best practices and responses toward safeguarding systems and applications. The scheme supports incident response and risk assessment programs and keeps security relevant while focusing on continuous visibility. A person that embraces the changing culture will activate security involvement through roles and responsibilities.

Active involvement is similar to igniting a switch. When keys are turned, or a button is pushed, the electrical flows, lights, engines, or power starts. Our ability to view data, watch television, or press the gas pedal should occur. Sometimes we must enable different standards, troubleshoot, or make emergency repairs, which activates the desired service.

In the technology arena, change can activate human firewalls. The concept requires the understanding of culture change and where their jobs align. The initial process involves management practices that integrate a buy-in and commitment to task objectives, which is the power switch. Once the button is pressed, there should be a sense of motivation, which is where people begin to engage in defensive thinking and system protection standards. Once the motivation phase executes, people place active involvement into actionable tasks. You can vision teams embracing the transition and securing applications, data, and assets. Active involvement is a constant effort and process that secures systems.

Active involvement is not an overnight transition. Long before executing changes, a corporation must implement strategies to motivate security involvement. The approach may consist of meetings and assessing performance, which provides insight into whether involvement exists. The existence can represent an indicator but not a direct answer on motivating security best practices. A better approach is to examine intrinsic and extrinsic motivational strategies. It requires people to activate security involvement through internal (intrinsic) or external (extrinsic) influence.

Intrinsic motivation allows people to approach security as defenders. You can envision defenders as human firewalls that embrace and think about cyber safety. There is a sense of internal reward and behavioral standard that naturally exists. Through the culture change, management may find this motivational type operating in third gear. When the word corporate transition exists, this type ignites an activation switch, which immediately starts the security engine. Extrinsic motivation requires external influence to shape involvement and behavior. A reward structure is needed to ignite defenses. The mindset is more of, "What's in this for me?" Through new cultural changes, this caliber of behavior may induce additional risks. There is a lapse in embracing the cultural transformation, and natural motivation is nonexistent. The activation switch and involvement require external influence—such as a day off or company award. Despite the motivational strategy being used, the goal is to embrace security changes. The intrinsic method is more straightforward and internally developed, while the extrinsic process is an exchange or reward that influences and changes behavior.

As stated before, the most beneficial path to active involvement is preplanning. When formalizing a staff, management must embrace and sell the corporate brand. The preplanning of actively involving staff members lowers risk and safeguards critical assets. The concept requires technical and non-technical teams to embrace cultural change. Picture corporations that individualize security best practices—the non-technical staff would have less motivation and involvement. The accountant or administrative staff may separate security when performing daily tasks. It gets even worse—data protection standards may fail, or the motivational switch may not operate.

Corporations spend countless hours reworking and planning changes. How staff members embrace and activate their motivational switch determines whether security adoption or adaption occurs. When a culture transformation executes, new technologies require adopting best practices, policies, and standards. Think about the transition between two security frameworks. If using NIST 800-53 was a standard and the new standard became proprietary—the best practices and regulations would adopt the new standard. The adaption model works entirely differently, as it requires a 100 percent buy-in to the cultural transformation. Its approach and mindset embrace the corporate brand. Picture emergent security situations that involve technical and non-technical teams. Each has a separate space, but the corporation will produce active involvement as a single practice through solid adaption. Risk management, incident response, or security management practices become its corporate norms. Each involved communicates, shares technical knowledge, and promotes the corporate governance strategy.

As new technologies surface and transformation occurs, there exists a constant adoption and adaption engagement. During the technology evolution, there have been many changes and transitions. The IT industry transitioned from desktop to mobile and from onsite to remote functions. Different standards and innovations created newer processes where technology embracement was at the forefront and integrated significant changes. The changes meant that technology alignment required engagement, integration, and more integration. One industry that benefited has been cloud adoption and adaptation.

Cloud integration brought upon a much different aspect and technical culture. Cloud customers had to embrace services that were typically locally owned but now are handled by cloud service providers (CSPs). The CSP's sole purpose is to sell technology as a subscription service and manage applications or hardware. The customer benefits because maintenance, upgrading, and ownership are a CSP's responsibility. Traditional local area networks (LANs) carried the same benefits, but the local organization assumed responsibility. Whenever a transition occurred, the adoption and adaption involved local IT systems and decisions. Cloud has introduced methodologies that require services allocated from various CSPs such as AWS, AZURE, or DISA MilCloud 2.0. Through these services, organizations had to adopt different service, operational, or privacy level agreements that are driven through data ownership and location.

The traditional method for data security involved internal data centers, applications, and infrastructure devices. These components created protective requirements, data authorization, and authentication mechanisms for access. The database administrator controls access to the internal databases, and networking teams make firewall policies to safeguard the data. The data security policies were created domestically, and policy adoption was organization-specific. Third-party or external policy considerations were nonexistent. Technical teams adapted to constant change, and protection standards brought upon newer management strategies, and then the cloud evolution created a significant transition, which the CSPs handled data protection.

A traditional data security scheme involves internal protection that locally manages visibility and security. Outsourcing clients' data to the cloud requires organizations to adopt new practices and standards. Since upstream and downstream liability exists, the cloud customer must create SLAs that protect their data, which requires audits, assessments, and visibility into the CSP's data security practices. Through a traditional LAN environment, data security was internally managed and visible. The cloud challenged security visibility, where organizations had paper-driven vs. external accessibility into data security status. Adapting to this change was challenging and risky, which is why SLAs exist.

Since the cloud is more massive than we imagine, data location is a concern. When internal environments exist, an organization knows where its data resides and how it's protected. Within the cloud, there are multi-tenant environments where data is stored. A multi-tenant environment is similar to an apartment complex. There are many tenants housed in a single building, and walls isolate each tenant. The cloud environment uses multi-tenancy in the same fashion, except each customer is isolated through security zones. These (walls) security zones provide safeguards and prevent unauthorized access. The tenants are housed in a data center with one hundred other customers. The modern method for data protection may include firewalls, isolation, restrictive rights, or encryption, which are traditional and bring different adaption standards.

When a cloud transition occurs, most organizations follow the CSP's data security standard and soon release some authority, which shows they have adapted. It's important to remember that security can never be entirely relinquished to the CSP. Organizations still bear some responsibility, which demonstrates adaption.

Embracing change is the evolution of many practices. Once it is fully embraced, an organization can view cloud technologies and behavioral influences and activate its security engine. The methods, processes, and programs are all synced and ready to secure data, where measurable outcomes become pivotal. It allows an organization to validate and understand security transition and culture change and what behavior or condition has succeeded.

Measurable tracking can benefit cybersecurity adaption. Before integrating a new change, an organization can have performance marks established. These marks evaluate the current security state and operational characteristics. The most visible identifier is behavior patterns, involvement, and risk posture. If these elements mirror from initial to completion, a corporation can successfully function and embrace adaptation. All the protective measures, safeguards, and security engines can motion to defend a "hacker's appetite." The corporation profile or branded image will demonstrate defense thinking and an inclusive culture that embraces change.

Cultural change can happen at any IT stage. As technology innovations occur, cybersecurity professionals must develop approaches that balance security and culture transitions. Legacy systems and management strategies are set to become obsolete, and behavioral techniques must include best-security practices. To instill and cultivate these practices require technical and non-technical delivery methods. In a professional scheme, we could highlight risks and security involvement. This may work in some sense, but balancing non-technical approaches and learning enhances the process. Here is where security awareness and the corporate brand intersect.

CHAPTER FOUR

BRANDING A TRAINING AND AWARENESS PROFILE

In 2005, I started my post-sea-duty career as a leadership facilitator. The duty consisted of delivering naval leadership methodologies, principles, and career progression knowledge to junior sailors. The initial training involved attending a training-the-trainer course in Norfolk, VA. The course provided initial entry into naval instructional delivery and curriculum development theory. The information and learning modules focused on naval leadership and development (NLDP) and practical tools to deliver effective training. Throughout the course, class members developed leadership training guides and publications and learned instructional techniques. During the middle of the training period, I began to organize teaching methodologies and their military value, which was an effort to align and brand naval leadership. To build memory and learning, our class created sailor-based information. The concept behind using sailor-based information was to help transition our experience into the classroom, and it's defined as a methodology that cultivates a sailor's training process and knowledge.

Upon graduating from the course, I received orders to Naval Station Mayport—the Center for Naval Leadership (CNL). There were signs, placards, and marketing materials in the classrooms, hallways, and instructor offices that highlighted the military's leadership philosophy and value. Students that entered the building were thoroughly trained and nurtured toward leadership and career development.

The CNL course modeled corporate-based leadership with a military twist, which existed because the curriculum integrated baseline leadership principles, and the objectives mirrored the private sector. The organizational approach separated the military and private sector philosophy and instituted a military training culture. There were two course modules that were referenced as the Workcenter Supervisor Leadership Course and the Leading Petty Officer Leadership Course. Each course supported different leadership progression requirements. The language, discussions, and group learning exercises integrated sailor-based information aligned to the Navy's mission and vision, and successful leadership practices.

The IT learning requirements mirror the naval leadership courses. IT professionals are required to receive professional training that aligns their skillsets and role responsibilities. The alignment integrates various concepts from lessons learned, previous attacks, technology innovation, or required skillset. As with the leadership courses, the lapse of branding would misalign the course content. The same exists for cybersecurity—so it's essential that the learning is branded and regularly evaluated, or security may not become secure!

Every IT organization shapes learning and awareness that mirrors individual career goals and cybersecurity skillsets. It provides an outlet to attain advanced knowledge, develop career requirements, and foster a security mindset. Corporations can utilize external vendors to build training packages or use training frameworks. Each is designed to deliver guidance and help shape the learning plans. If we Google the internet and type in "training framework," there would exist many web pages, and deciding which is beneficial is challenging. So how can a functional framework exist that brands the corporate image? There is no complete answer, but the framework must integrate security gap closure and skillset analysis.

A framework is somewhat similar to a picture frame. You have all the sides and borders attached, which regulates the overall dimension or picture size. If you increase the frame size, the image expands. When organizations brand their training framework, the security needs, elements, training, practices, goals, and defensive concepts represent the cultural image. As new requirements surface, the organization reinvents its framework and examines new learning methodologies that brand the culture, which expands the training and learning points. Once it's achieved, the framework becomes the ideal model until the culture change. Alternatively, the outcome is distinctive and caters to training features that model the security program objectives.

Security learning and awareness have had their challenges. Every year, organizations build and promote annual cybersecurity awareness training. The courses range from fundamental to advanced education, and the modules support the organization's knowledge base and technology goals. When developing the course modules, the corporate technology culture and security history must be included. Each provides security deficiencies and skillset gaps and helps establish the learning framework. For instance, an organization that uses biometric scanning should incorporate biometric authentication as a training module. The same concept applies concerning incidents, data breaches, and onboarding requirements. Suppose a corporation tracked events and security violations. In that case, it could provide direct knowledge on training needs, security training objectives, and skillset development.

Another challenge for security learning and awareness is employee engagement. If employees learn boring, redundant, or useless information, corporations can achieve 100 percent completion; but does the culture buy-in exist, and are they thinking security? A more formative concept is to include realistic cybersecurity information. This is similar to sailor-based training in that it's attractive and relatable, and it cultivates the organizational mindset. Generic training can work, but does it support the training framework? Corporate-based learning models include security indicators and realistic training that address previous attacks, weaknesses, or violations, and it develops the necessary risk remediation skillsets.

While working on several DoD contracts, our team trained on many security products. These products supported the security approach, vulnerability remediation, or defensive tactics. We received military-aligned training compared to previous staff members, which meant the learning mirrored the Navy security, risk, and technology profile. The outcome was a significant advantage since prior security teams experienced substantial gaps in their training approach. One particular product was McAfee ePolicy Orchestrator—a security tool that provides real-time visibility and security event analysis for desktops. The last team received McAfee ePolicy Orchestrator training in a classroom-based environment, which omitted the Navy's security requirements. For instance, the classification for military systems is secret and requires privileged access—the original staff trained on generic access and security response procedures. As the project developed, the previous staff had to receive additional training, which was time-consuming. Our team conducted onsite training that modeled the Navy's security vision and skillset requirements. The training consisted of an onsite McAfee ePolicy Orchestrator SME's demonstration classroom and technical theory. We had the opportunity to integrate the Navy's security standards and receive branded learning, which reduced additional training requirements. The approach demonstrated for McAfee ePolicy Orchestrator is just one example of culture-based training. If we dissect the technology industry, different techniques are used to brand skillsets and cultivate learning, and most are straightforward, while others require tailoring.

The term tailoring means the training is branded and constructed to support the Navy's needs.

While developing leadership training, I experienced various branding techniques. One particular method was to attract students' interest. The culture for today's industry utilizes games and social media as learning channels. Organizations can incorporate these same principles to attract attention. For instance, company "ABC" may create a corporate-based social media page with various security issues. The page can have different names that relate to their departments and divisions. The outcome would include data breaches, information sharing, or phishing. As employees navigate the page, they can engage in relatable information that coincides with job assignments. Another option is to incorporate scenarios based on recent data security violations or security incidents and display monetary signs when correctly answering questions or identifying issues. When either possibility exists, a corporation can model a corporate training framework, fine-tune its employees' security engine, accelerate cultural knowledge, and prevent security incidents.

It's been said that employees are the core enablers and fault causers when incidents occur. Corporations face minimal issues such as password complexity to unauthorize access. It's known that awareness training occurs annually, and still, corporations suffer attacks and data breaches. So the question arises: how can we build a culture that thinks cyber? One particular step is training through end-state readiness. The concept implies that the security training's end goal is to develop security involvement and promote cyber safety. Another methodology organizations can consider is to evaluate the trainer. It's impossible to have a successful learning and awareness program without proper instruction, and every training culture brands its instructors to sell its vision. Having a disqualified instructor-based program can induce risks and further complicate security goals. Picture a security awareness program where an instructor exhibits a negative attitude toward the course. How can the employee's security engine start? After all, the instructor has the keys. When businesses brand, instructor-based teaching programs must receive considerable attention. If the instructor fails, so do the training programs, which create security errors and technology risks.

Branding training exists outside security awareness. There are skillset development, learning, and certification pipelines to support the corporate training framework. Each year, security stewards pursue certifications such as the Certified Information Systems Security Professional (CISSP), Certified Ethical Hacker (CEH), or the Global Information Assurance Certification (GIAC). These are just a few IT certifications that hold value and a pathway to sustaining employment. The pursuit can be very stressful and costly. The average price to attend a CISSP course plus the examination fee is about $3,500. When examining true-skillset requirements, the certifications are valuable. The CISSP depth of information focuses on the major security domains and relatable concepts toward the security market. The relatable training consists of cloud technologies, networking, attack profiles, risk mitigation, or legal ramifications for data breaches.

When developing a security stewards' training profile and goals, it's imperative the technology culture, incidents, and skillset analysis are included. For instance, an organization that develops software code may wish to certify their employees on the Certified Ethical Hacker (CEH) pipeline. The CEH covers application security vulnerabilities and penetration testing requirements. The certification provides knowledge on removing software bugs and techniques hackers employ to break application security defenses. For more senior-level software coders and developers, an organization can consider the Certified Secure Software Lifecycle Professional (CSSLP). The CSSLP provides advanced knowledge and concepts to deliver software products using the Software Development Lifecycle (SDLC). Candidates learn how to develop and test software code before product delivery and learn to incorporate security practices into the SDLC. Despite which certification an organization pursues, the path must align and model their security profile and defensive mindset: *what are we protecting, and how do we navigate skillsets?*

Skillset navigation is driven through various security gaps, weaknesses, incidents, or technology innovations. When organizations evaluate their annual training goals, incorporating previous events and attacks determines skillset requirements. The outcome identifies various skillset deficiencies and how they affect the technology culture. When reviewing previous attacks or incidents, an organization may see trends in network attacks and skillsets protecting network infrastructure devices such as firewalls, switches, or routers. The skillset alignment provides the tasks required, knowledge level, or functional role to secure the network perimeter devices. Offering certifications such as SANS GCIA, GCSA, or GPEN would be an aligned certification for protecting the devices. Another option is to implement on-the-job training (OJT)—commonly called hands-on training.

Integrating an OJT program upgrades an employee's skillset and cultivates security. When developing an OJT program, employees are trained onsite and provided relatable information about the corporation, working tasks, and roles. Including incidents and attacks are provided to highlight the technology culture weakness and required growth. As previously discussed, a growth mindset is an improvement process. When including the prior history and focused training modules, an organization provides direct security knowledge and safeguards information assets. In the example of network security protection, a security steward could receive hands-on and organization-specific training. The training would reinforce defensive schemes, security configuration, firewall rule structures, or network authentication standards. Through the OJT program, the employee can better address future network security attacks stemming from real-time, functional, and organization-specific training.

The value of OJT extends to application-based learning. As previously discussed, the OJT program models the security culture and integrates aligned or organization-specific training. Application-based knowledge is a technique where the learned information applies beyond the training environment. When incidents surface and attacks need a quick response, application-based learning must interact. Through the process, security mitigation and perimeter defense can evade further intrusions. Imagine working in a SOC environment where every intrusion or attack response and mitigation activity lasted about one minute. The entire process would require well-trained, security-focused individuals that understood attack scenarios, defense strategies, and an organization technology architect. The root cause for a quick response relates to the application-based learning principle: *trained with relatable information and provide real-time training objectives.* If adequately employed, the OJT and application-based learning technique can develop strong cybersecurity skillsets, enable career development, and model organization-specific training.

There exist challenges with an OJT program. Trying to deliver relatable training can occur, but at what percentage? Organizations can pinpoint various training aspects and model real-world situations. However, for some odd reason, some security tasks are overlooked, which results from generic training. OJT is designed to model real-world and practical problems involving a trainee; it's more beneficial to provide OJT when high-tempo operations exist. During this stage, every known task and security requirement surfaces. The knowledge gained is paramount, and their proficiency grows due to the repetitive work. As the trainee becomes more engaged in the OJT program, their skillsets sharpen.

Skillset development, branding, and learning have many extensions and interactions that exist outside their functional roles. It can exist as a monetary exchange, while others are employee-defined. Monetary defines the costs and incentives received through a training framework and its outcomes—such as the required CISSP training costing $3,500. When designing the return on investment, the $3,500 adds more value and shapes the security brand since it aligns with cybersecurity. Employee defined addresses how a security steward benefits the organization's training needs and their interaction. One particular interaction is with the Employee Value Proposition (EVP).

EVP is an offering provided by an organization in return for employee skillsets, capabilities, and experiences. The EVP is an employee-centered approach that aligns with existing skillset requirements. Employees who complete skillset-based training bring valuable solutions and computer network defense (CND) tactics to safeguard the most critical systems. Organizations can successfully sustain top talent. The employee demonstrates a more branded approach and understands corporate security challenges. In return, an organization can offer pay incentives, cash recognition, additional training, or employment security. When EVP achieves, corporations can minimize risks, structure defense capabilities, and align job roles.

To accurately demonstrate EVP, businesses must understand solution principles. Solutions are where they resolve issues and provide maturity. Growth is significant in the security defense system because it helps to secure enterprises and develop skillsets. The entire process requires solution frameworks that make risk mitigation successful.

A solution framework is a formative process that enables resolutions. Once skillsets are developed and employed, the security steward can resolve the most sophisticated security challenges and safeguard applications, systems, and resources. Without prior knowledge and skillsets, intrusions' growth would overshadow technology, and hackers would win the war. When coupling the EVP and skillset development, we can develop defense tactics and provide career progression. Each employee would express motivation and security thinking far ahead of incidents. Typically, an environment may hear, "What's in this for me?" Through EVP, a security steward knows the values and requirements to project security involvement.

When incidents occur, the security steward ignites their security engine and thinks protection. Every organization wants employees to think along similar practices and promote a cyber-safe environment, which can succeed when value-added incentives exist.

In 2011, I started pursuing the CISSP certification. After spending hours watching Shon Harris videos and taking practice exams, I passed the examination in 2012. Immediately after hearing the good news, I updated my resume and began a new security journey. One goal was to utilize the certification as a value-added incentive and advanced security role, which meant an organization could "brag" about CISSP-qualified employees. In October of 2012, I received a call concerning new employment opportunities.

The recruiter expressed that having my skillsets would resolve many security issues. He later explained that the position wanted someone that "lived" like a CISSP, which meant a person preached and lived security. I accepted the job offer, and after about three days on the project, I received my initial task. The objective was to remediate 10K vulnerabilities across Unix, Windows, and third-party applications. As I engaged the journey, risk management's thought surfaced, which was the first CISSP domain? The domain discussed risk management, security policies, mitigation methodologies, and the risk analysis process. Here is where the relationship between skillsets and solutions became valuable. Knowing vulnerability management and technology was centric in developing the solutions, I was able to deploy a solution practice that mitigated risks. For instance, the certification discussed prioritizing severities and remediating the most critical vulnerabilities first. Since the tasks required risk knowledge, I was knowledgeable and had the proper skillsets. Throughout the lesson, I heard Shon Harris's voice and risk management topics hardcoded into my brain cells. The point is that the CISSP certification is aligned to a critical skillset. Had I pursued other IT certifications, the entire recruitment and hiring process would have never existed.

Once skillsets and awareness profiles become ongoing, an organization must build a community of practice where learning programs and delivery methods continuously operate. The continuous process exhibit features where knowledge gained and learning models the corporate security approach. Through various tactics of cultivating awareness, technology professionals can interpret security requirements and mimic human firewalls. The process requires a complete transition into active thinking and involvement and understanding organizational value. Once the value chain starts, safeguarding information assets becomes a practice and security professionals enable their cybersafe skillsets. The cycle continues for new employees, which keeps the method active.

Security awareness and training can transition the culture security model into ongoing work. As stated throughout the chapter, organizations must align and continuously evaluate their learning programs and their contents. The information developed and framework utilized must be branded and carry the corporate security image. Over time the image may change and require adjustments and new processes, but constant involvement will never change. Trying to instill ongoing participation is more mental than technical. People are motivated differently, and EVP may operate on behalf of employee A. However, employee B may require different EVP requirements. As a leader that drives inclusive cultures and promotes learning, it's imperative to brand. The end-state develops security and relatable knowledge as a single entity. *Teach the student and not the test!* Now, let's move ahead and utilize our branded mentality and also see what practical tools are available.

CHAPTER FIVE

INCLUSIVE CULTURE TOOLKIT

The past chapters described various methodologies concerning inclusive culture development and practices. Undoubtedly, an inclusive culture develops organizations and aligns the corporate brand, growth, behavior patterns, and buy-in structure. Survivability depends on modeling a successful strategic plan. The model may resemble project management theories and practices or just manageable tasks. In either case, the operation requires useful toolsets.

Toolsets can be defined as a set of practices, processes, or avenues to navigate an inclusive culture. Every security environment uses tools, and they can be technically or program-based. When relating the inclusive culture and toolsets, the discussions address program management strategies. These methodologies produce a series of steps and processes that support cultural development.

When deploying toolsets, businesses must define their usage. It's no different than mechanical tools in that they serve a specific purpose. For instance, an organization that employs skillset development programs may use the technology culture and metrics to understand training needs. Through using metrics, they can identify security trends that require attention and whether the tool functions.

There are numerous approaches and ideas to form toolkits, such as extracting random thoughts or strategizing usage. Randomly selecting is a "shooting from the hip" concept because the tools function under trial or error. Corporations waste time, resources, and labor following this strategy. A branded tool, a Cybersecurity Mindset, and an inclusive culture are more beneficial. They allow corporations to utilize unique programs, steps, and processes when designing a toolkit and strategize the toolset purpose. A tool's purpose would help develop cybersecurity within the inclusive culture through stages, lifecycles, checklists, or practical steps. When correctly used, a business can scope and deploy useful tools that align with its brand and improve security response and coordination efforts.

Now, let's explore some practical tools designed to build an inclusive culture. The implementation selected for our example can model many real-world examples. For instance, a business could perform data collection, execute checklists, and deploy mitigation practices. The list may contain relatable checks that mirror the corporate image. In the example below, the model provides a lifecycle process that operates continuously. We will be using *Standards, Impact, Program Defect, and Potential Damage* to understand the model.

- **Standards**—These are defined as the primary operational principles. They provide various practices and images the inclusive culture should mirror and practical ways related to security and technology development.
- **Impact**—The impact phase provides an overall rating for failing or dysfunctional standards. The outcome indicates whether the inclusive culture standards are Exist, Semi-Exist, or Required.
- **Program Defect**—The term describes an incorrect behavioral pattern or a failing or dysfunctional standard. The outcome indicates that a standard activation is required.
- **Potential Damage**—When program defects occur, potential damage can make use of the error. The potential

damage can be behavioral patterns, non-activated standards, or a failing security engine.

Phase I: Analyze the Current Impact

The purpose of Analyzing the Current Impact is to implement business security standards. The programs, practices, behavioral traits, and defensive mindset are evaluated and standardized. The effort starts with understanding the cultural identity and its evaluation procedures. Business ownership, employee security engagement, or security profiling helps to define cultural identity.

The initial stage requires observation into behavioral patterns, security engines, active motivational practices, and a buy-in structure. Observers may inject statements through real-world scenarios or working with security personnel. The outcome describes the cultural boundary and security roles and responsibilities. Each working task or objective receives a comprehensive examination and deep-dive analysis. The observer can inject scenarios and taskers to determine outcomes or build extensive tests—such as parallel observations. The concept uses multiple observations with the same input on different systems simultaneously to reduce observational time. It evaluates a culture and its active and reactive responses. For instance, there may exist events that jeopardize the security program. The security team may never experience changes or certain events through normal operations, so invoking an intrusion attack on a nonproduction environment provides the observable data. The attack outcome and its impact can determine the rating (Required, Semi-Exist, or Exist).

Rating	Definition
1. Required	The security feature is required, and its outcome will have minimal impact on the security culture. The culture can function with or without the security feature.

2. Semi-Exist	The security feature must exist, and its outcome will have a substantial impact on the security culture. The culture can partially function without the security feature.
3. Exist	The security feature must exist, and its outcome will have an operational impact on the security culture. The culture cannot function without the security feature.

TABLE 1. RATING DEFINITIONS

An organization's security profile can incorporate various or distinct features. They resemble security classification schemes in that the culture is labeled and profiled. Each label relates to the functionality that uses security—such as financial, health, IT, accounting, or human resource. Each functionality uses security for various reasons. Also, each security feature has a standard rating that relates to data classification.

Now let's move deeper into conducting a precise impact analysis. In our scenario, we will utilize a security feature and evaluate its rating. The procedure can be tailored by changing the functionality and security features. For this example, we will use the HR department.

Security Functionality	Security Feature	Rating	Score
Semais.1	Culture Training	Semi-Exist	2
Semais.2	Behavioral Influence	Semi-Exist	2
Semais.3	Growth Mindset	Exist	3
Semais.4	Active Involvement	Exist	3
Semais.5	Buy-In Structure	Semi-Exist	2
Semais.6	Value Proposition	Required	1

TABLE 2. EXAMPLE OF CULTURE RATING

KEY FACTORS

- Overall Rating: Score (Total) / Score (Line Items) and Round-Up or Down
- In our example: 13/6 = 2.17 or 2, which equates to Semi-Exist
- The Standard equals "Semi-Exist."

As indicated, the security culture operates in a Semi-Exist state. The outcome does not define the organization as dysfunctional or holding significant defects, but it does represent the culture rating. Businesses can use the score to determine standards and then identify their implementation procedures. Let's examine a few examples.

Table 2 provided detailed rating procedures for the security culture and the security functionality aligned with the security features. In our next step, we will expand the security functionality into standards, which are low-level examples that describe the culture operations. In a functional setting, the standards support security involvement and evaluations and guide the cultural development process. The first task is to develop standards and define their purpose—as described in **Table 3**.

Security Functionality	Security Feature	Rating	Defined Purpose
Semais.1	Culture Training		
Standard	Skill Set Alignment	2	Communicates skillset alignment
Standard	IT Certifications	2	Aligns critical technology skills
Standard	Security Awareness	2	Provides adequate skills
Standard	Scheduled By HR	2	HR schedules training
Standard	Training Outside of Role	2	Complete training for other areas

Semais.2	Behavioral Influence		
Standard	Management Controls	2	Manages cultural behavior
Standard	Response	2	Thinks cybersecurity
Semais.3	Growth Mindset		
Standard	Measure Progress	2	Make decisions about progress
Standard	Cultural identity	2	Drives the direction for growth
Semais.4	Active Involvement		
Standard	Security Engine	2	Operate through motivation
Standard	Measured Through Assessment	2	Only an assessment can provide measurements
Semais.5	Buy-In Structure		
Standard	Implemented	2	Continuous viewed by management
Standard	Management Responsibility	2	Designed for Management
Semais.6	Value Proposition		
Standard	Security Defense	2	Employee gain skillsets

TABLE 3. EXAMPLE OF STANDARDS

Each standard provides the operational characteristics and rating designation. **Table 3** was developed to describe how security functionality, features, and standards co-exist. In a healthy environment, a corporation may have additional line items. Many security functionalities and features could exist depending on the culture, technology concerns, and corporate structure. The next phase will involve branding the standards, which is dependent on the security architect.

Phase II: Architect and Brand Strategies

The security features build cultural identities and their brand. It's relatable to the System Development Lifecycle (SDLC), which provides an architecture where features and identities match. Aligning the cultural identity and standards is critical. These are the core enablers that shape toolkits and branding. Picture a situation where the security feature received a "Semi-Exist" rating, but "Exist" was selected during the selection process. How would the cultural identity align? Also, would there exist program defects and potential damages between the corporate mission and employee security engagement? These are pondering questions to consider when scoring and evaluations are misaligned. Now, let's move ahead and expand into branding. Here are a few notes:

- Stay consistent with the defined and tailored standards.
- Creating a specific deviation may require a new "Culture Rating."
- Start with the image in mind—that means the culture-specific information is vital.

We selected "Semi-Exist" as the cultural standard, which means Table 3 must be tailored. Since business environments change periodically and new technologies are onboarded, corporations must re-evaluate their framework and tailor specific standards. Another example that describes tailoring is modeling or customizing, which means culture-specific standards are selected. These culture-specific standards describe the exact security requirements and their functions. In our example, we would tailor all the "Semi-Exist" standards and then specifically deploy tailoring. In the end, some of the "Semi-Exist" standards may not apply, which results from technology or security engagement procedures.

Security Functionality	Security Feature	Rating	Defined Purpose
Semais.1	Culture Training		
Standard	Skill Set Alignment	2	Communicates skillset alignment
Standard	IT Certifications	2	Aligns critical technology skills
Standard	Security Awareness	2	Provides adequate skills
Semais.2	Behavioral Influence		
Standard	Management Controls	2	Manages culture behavior
Standard	Response	2	Thinks cybersecurity
Semais.3	Growth Mindset		
Standard	Measure Progress	2	Make decisions about progress
Standard	Cultural identity	2	Drives the direction for growth
Semais.4	Active Involvement		
Standard	Security Engine	2	Operate through motivation
Semais.5	Buy-In Structure		
Standard	Implemented	2	Continuous viewed by management
Semais.6	Value Proposition		
Standard	Security Defense	2	Employee gain skillsets

TABLE 4. EXAMPLE OF TAILORED STANDARDS

The next task consists of executing a deep-dive analysis. The process requires advanced observation or research into security identity. Evaluators may examine specific security engagement procedures and cultural responses to define standards further. Some of the selected standards may be removed or added. For instance, the "buy-in structure" may be described under "growth mindset"—so excluding the Semais.5 security functionality would suffice. Now that the standards are selected, let's deploy the measures into a cultural setting.

Phase III: Deploy the Strategies

This phase relates to an organization's mission and business objectives. It focuses on cultural identity and branding within the technology community. It somewhat describes transitions and involvement, which requires employees to embrace change. To energize both methodologies, observers must heavily engage the security team and climate. The end-state is to produce a culture that projects the brand. A starting point is to extract and deploy the "Tailored Standards."

When implementing the tailored standard, the observer may involve technology standards and security engagement. Their purpose benefits culture branding and its dependencies. If we evaluate *Behavioral: Response—Think Cybersecurity*, the end-state would produce an environment where Semais employees' mental state exhibits data protection and security onboarding. Managers may enforce strict behavioral standards that force security thinking. The entire process onboards Behavioral: Management, which means management drives behavioral involvement. Now that the security states are deployed, the next stage will require an in-depth assessment.

Phase IV: Assess the Inclusive Culture

An assessment provides two indicators: the security culture alignment and its operational characteristics. The current employee engagement toward defending organization assets and information is core—the entire phase onboards the inclusive philosophy, cultural identity, and the corporate security state. Involving exclusive thinking can cause program defects. For example, an observer may choose Behavioral:

Response and notice the security teams operate independently, which means their reactions are non-cultural based or exclusive. Further analysis may reveal employee engagement and security contribution as the root cause. The learning point: deploying a cultural assessment can measure employee involvement and active engagement.

In Phase II, we defined the culture standard. In this phase, we will determine whether the security engine operates and what program defects exist. In our example, the specifications would be security standards. The goal is to evaluate the standard and security events.

As previously discussed, a corporation can have many security functionalities and features that derive from its security culture and operational characteristics. The security culture may include several security teams that manage incidents, network configuration, data privacy, or events. When executing an evaluation, it's risky to evaluate every security area. So, a viable solution is to sample datasets or assess areas that require critical attention. **Table 5** outlines a typical dataset, and **Table 6** provides the evaluation categories.

Evaluation Area	Sample Data
Skillset Alignment	• Sample 10% of security employees • Sample 10% of the required certifications
Response	• Evaluate 3 of 10 security engagements for incidents • Evaluate 2 of 12 employees' motivational approach
Cultural Identity	• Sample 25% of security employees mental approach • Sample 10% of the corporate executives' behavioral influence strategies • Sample 10% of employees' engagement to growth
Security Defense	• Sample 10% of employees' active involvement practices • Evaluate 10% of human firewall practices • Evaluate 5 of 10 security engagements for incidents • Evaluate 30% of employees on the cyber thinking capacity

TABLE 5. SAMPLING DATA

Technique	Capabilities
Observe Mental Approach	• Discovers active involvement in practices • Identifies cultural inclusiveness
Verify Security Engagement	• Discovers whether security response operates • Determines whether a mental approach exists • Determines whether the business has a growth or fixed mindset
Observe Security Teams	• Determines whether they think cybersecurity • Determines their behavioral patterns • Determines whether they can become human firewalls

TABLE 6. EVALUATION CATEGORIES

Conducting the Cultural Evaluation

1. **Pre-Assessment**—For this step, start by gathering the system and assessment information and determine what culture-specific data to evaluate. The information provides the assessment structure and evaluation schedules. Next, define the sampled dataset and security standards. Last, develop the assessment approach or evaluation categories. **Reference:** Table 5. Sampling Data and Table 6. Evaluation Categories.

2. **Execution of Assessment**—Schedule and execute the assessment by using **Table 6**. The results provide indicators and program defects that exist and their potential damage. The current security culture and standards receive a complete evaluation. A typical evaluation may resemble the following:

Evaluation Area	Security Feature	Sample Data	Evaluation Category	Security Results	Potential Damage Score
HR.6	Security Defense	30 of 100 employees selected	Observe the Security Team and determine whether they think cybersecurity.	After conducting the cultural assessment for security defense, the evaluation revealed that 20 of the 30 subjects exhibited cybersecurity thinking.	67%

TABLE 7. EVALUATION PLAN

3. **Post Assessment**—During this phase, the results, root-cause analysis, and mitigation recommendations are assessed. The potential damages are mitigated or transitioned to a corrective action plan (CAP). Corporations examine why, how, what, and when the potential damage surfaced.

Evaluation Area	Security Feature	Security Results	Potential Damage Score	Corrective Action Plan	Date Opened	Date Closed
HR.6	Security Defense	After conducting the cultural assessment for security defense, the evaluation revealed that 20 of the 30 subjects exhibited cybersecurity thinking.	67%	The business will implement training standards that focus on cybersecurity thinking. Each employee will receive a follow-up evaluation.	Dec. 23, 2019	Jan. 23, 2019

TABLE 8. CORRECTIVE ACTION PLAN

Once all the data is collected, the business can track progress and make additional recommendations for the standards. The CAP) is continuously updated and reviewed until the damages become acceptable. The acceptability and follow-on-action items execute within the next stage.

Phase V: Evaluate and Make Transitions

When making transitions for the business, the organization examines the program defects and CAP. The traditional cultural identity may transition into a more inclusive environment. The program defects and exclusive behavioral patterns are realigned. Making the process functional requires an in-depth CAP evaluation and transition plan.

The CAP evaluation is a deep-dive analysis of the program defects and its treatment plan. The organization evaluates the culture-specific program defect and determines its potential damage and exposure. In our example, the security defense is potential damage, which drives security decision-makers in examining its overall cultural impact. If the security feature poses potential damage, the business can institute emergency remediation. Its sole purpose is to transition potential damages into a remediated state and enable security. Once the procedure activates, culture standards become operational and inclusive.

The traditional method does not involve emergency remediation. In this format, a corporation may openly operate. A designated security manager must review the CAP and ensure its impact provides minimal disruptions. For instance, a security steward's cyber thinking capacity may produce the following: Critical Task: Passed Non-Critical Task: Failed. Under these circumstances, a security manager can execute "Operations," which means the potential damage posed minimal impact.

Once the culture becomes operational, the organization can evolve its security visibility and continuous monitoring programs.

The image projected shows the culture embraces cybersecurity and thinks defense. The complete transition requires constant observation and active security engagement, which is the last tool!

Phase VI: Perform Continuous Observation

The definition of continuous observation requires active security engagement, motivational strategies, and human firewalls. These pillars enable visibility into ongoing defense practices, security involvement, and response procedures. It's a complete security engine that operates in a continuous and observable state.

Monitoring the security culture's mental approach is a constant practice. In our example, the security features are considered working parts of the security engine, and sustaining their visibility requires specialized tools. The previous tools focused on evolving and assessing the culture, but this phase transitions into ongoing work. The two most critical tools are reassessment and interim checks.

A reassessment can occur ad-hoc or through scheduling. It requires a partial assessment based on Phase I to Phase V and incorporates the same security standards and features. For instance, a business may decide to evaluate security defense. The previous method was more formal since there existed five stages. Through the informal approach, the observer can monitor cybersecurity thinking reports, incidents, and communication practices. The benefit of informal assessments is that it presents real-time observation. Scheduling may induce formal observations that force operations and cultural involvement. While the outcome is positive, it still does not guarantee ongoing engagement. Place this in your toolkit!

The use of interim checks is similar to a reassessment, except it's organized. The security and cultural development stages receive a planned monitoring service. An organization can choose various standards and periodically check its status. In a pure project-based organization, you may hear terms such as data calls or action items. A data call requires data collection or information gathering. For instance, a data call can request a list of aligned certifications and compare the results against a required dataset. An action item requires completing a task. In our example, it may require passing a certification test. When the ongoing work executes and finishes, the corporation remains vigilant and promotes cultural identity.

The toolkit provided various assessment examples and ideas. The information is vital toward understanding inclusive principles and security operations. The proper usage varies between organizations, and there is flexibility in using the model to introduce culture-specific ideas or just use certain phases.

Cybersecurity engagements must remain an ongoing and continuous practice where corporations grow and develop security. Every culture embodies challenges and transitions. Working harder is somewhat a loose and overused term. To bridge security defense and culture involvement is a skill-development process—where operations and technology approaches function as one, and security professionals operate smart security practices.

Smart security is more significant than its meaning. The entire security architect requires environmental knowledge. There are different scenarios, practices, and ever-changing processes within the technology space. The experience and "thinking cyber outside the box" become critical when deploying a Cybersecurity Mindset. The security vision and protection schemes must match and promote the cultures' baseline knowledge of the best security practices. How the security culture approaches the challenge determines success or failure and whether situational awareness survives.

• VIRTUALIZED PATH •

- • Inclusive Culture
- • Situational Awareness
- • Risk-Based Thinking

SITUATIONAL AWARENESS

In 1995, the US Navy started to consolidate similar job classifications. Many of the technical ratings merged, and so did the sailors. It was a transition period driven by system changes and budget constraints. For instance, the Data System Technician (DS) career field was absorbed into the Firecontrolman (FC) classification. The DSes maintained tactical and mainframe computer systems, which were used to support combat detection and missile defense. The FCs deployed missiles and various weapons at enemy targets and sustained similar shipboard systems. During the transition, I received orders to the USS *Spruance* DD-963. The *Spruance* was a combat-driven ship that provided aircraft carrier defense, anti-submarine warfare, and enemy engagements.

The *Spruance* transition was a tense situation since prior skillsets were not 100 percent FC-related. Even today, the first twenty-four hours onboard the *Spruance* remains a learning lesson concerning different cultures. Just imagine being new and learning the ship's culture engagement while high expectations exist. After about thirty days onboard, I felt like a combat-driven sailor. The language, command vision, and crew relationships became embedded.

The combat-driven attitude was tested as well. During a divisional meeting, our division officer paused everyone to communicate a critical announcement. Usually, his pause signaled a task needed immediate attention. For this day, it was pretty different. He initiated the conversation by stating, "The ship needs a highly motivated and combat-driven sailor to assume the air-defense watch. The watch requires a senior-level sailor from the FC division." Immediately, everyone looked around and said, "Dewayne, it's your watch." The division officer looked as well and stated, "You know you were going to be selected." There was nothing to say except, "What watch?"

After the divisional meeting, the division officer provided additional details. Within a week, I assumed the air-defense watch in the ship's Combat Information Center (CIC). The CIC was where combat decisions, missile deployment, tactical operations, and warfighting exercises execute. Within thirty days of watchstanding, the combat defense mentality surfaced. There were moments when tactical decisions and operational knowledge required strict attention to detail and accurate communication. The captain and senior officers always asked questions concerning aircraft vicinity and origins and whether the ship's threat posture was safe. Luckily enough, I communicated and provided data for the ship's threat environment and aircraft's origins. Today, I believe the knowledge gained and information sharing relied upon environmental information and intelligence gathering. Studying the naval combat policies and enemy aircraft made the watch successful, and here is where situational awareness surfaced.

Situational awareness (SA) extends into many labor categories, personal involvement, or working groups. The law enforcement industry practices the use of situational awareness when approaching or stopping a vehicle. They use baseline information to understand the driver, vehicle, or current environment. For instance, the baseline information could simply mean, "Did you previously stop this car or have any amber alerts?" Obtaining the data helps in projecting specific actions, such as calling backup or building safety zones.

The technology space has used situational awareness to stay ahead of threats, weaknesses, and vulnerable situations. In some formality, IT personnel gain knowledge just like the law enforcement and respond and engage security. It's known that every problem or incident poses the responder to enact and think with reasonable decisions, and proper SA cannot guarantee good decision performance. There is a problematic chance that error will occur because of an inappropriate response. Furthermore, SA is influenced by the information process, which means that SA must continually adapt. The adaption is based upon the environmental conditions and changing states. Learning what variables are persistent and relatable is a constant reminder that SA is ever-changing, and so should our alignment to the environment.

CHAPTER SIX

ENVIRONMENTAL KNOWLEDGE

Situation awareness defines our environmental knowledge and incident response procedures. If a condition or process operates abnormally, a potential problem may exist. When exiting our homes, we hold knowledge concerning furniture arrangements and objects. Upon reentering our homes, we naturally assess both by default. These are our indicators and sensors, which are typically deployed and can indicate abnormal arrangements and conditions.

Another element that defines situational awareness is threat conditions and responses. The threat factor extends into safety and securing the environment, health, or state. For instance, military commanders assess the threat impact upon engaging an enemy target before responding or deploying additional forces, and the naval warfare strategies have embodied the same concept. Its combat-driven ships operate through threat postures and readiness conditions. If the environment poses unsafe conditions, the commanding officer may respond by executing warfare engagements. The response level and reactions dictate the ship's survivability and warfare success, which means human lives, ships' conditions, or mission goals still exist.

Situational awareness survivability can exist in the military, corporate America, law enforcement industry, or personal lives. Its success depends upon several critical drivers:

- What we know and understand serves as repositories.
- Every situation has a definitive meaning and normalcy.
- Maintaining alert requires human sensory: eyes, ears, nose, and touch.
- The threat hazards or impact drives our responses.

Cybersecurity and SA operate collectively in securing information assets and data. The management of security threats requires incident response procedures. The response level is based upon the threat state, such as ransomware or network intrusion. Most security organizations carry an incident response plan that outlines threat conditions and response procedures. When integrating SA, the security culture thinks about the affected systems or technology and the present indicators. For instance, a SOC employee may see an amber or red alert that signals abnormal activity. Immediately, their mindset shifts from regular operation to the situational response state—defined as the most successful reactions to specific situations. In the incident response plan (IRP), procedures may exist that address amber or red alerts. The SOC employee uses the baseline information in understanding reactions or solutions.

Another example is that the SOC employee can have previous knowledge concerning a system state. The information is gained through understanding the security culture, past response procedures, and IT operations. For instance, they utilize the corporate Security Information and Event Management (SIEM) tool to understand its operational security culture and intelligence platform. Since the employee uses the SIEM daily, their SA responds in seconds when abnormal events occur. Typical unusual procedures are invalid login attempts, rogue assets, blocked data traffic, or unnamed computers having system access. Without baseline knowledge, the analyst would react slower and could advertently increase the risk. The entire intelligence processing, system culture or operations, and the analyst response procedures collectively stem into environmental knowledge.

In 2009, the US Navy consolidated its networks under the Host-Based System Security System (HBSS) application—the official name was McAfee ePolicy Orchestrator (ePO). The application's role was to monitor desktops' intrusive activity and events. Typical tasks included reviewing various logs, events, software changes, and rogue information, and the operations analyst would perform analysis on events and submit incident reports concerning cybersecurity intrusions or potential system violations. While on the project, my awareness was raised concerning the security environment and the regular threat activity. When standing the watches, information concerning threat feeds and data traffic became a norm. As time progressed, the standard indications became the baseline knowledge or mental database and represented the system's characteristics. Through the different tasks, and when situations differed, my defensive thinking surfaced, and so did the human firewall.

Environmental knowledge transcends several security functions. Since security operates within administrative, technical, and operational lifecycles, so does environmental knowledge. The underline definition also extends into a cultural image and active security involvement. As previously discussed, these pillars promote the corporate brand. The demonstration of environmental knowledge operates the same—since SA depends upon the pillars.

Gaining cultural experience can broaden SA and promote better response. Engagements and decision-making require prior experience or learning involvement. A typical security practitioner that engages in the corporate culture learns security operations and incident response procedures, which is not an overnight success. After involving time and assuming tasks, the security practitioner "learns the system." The system consists of cultural practices, system functionalities, or prior incidents.

The SA learning curve exists beyond the SOC environment or security team tasks. The strategic teams and CISO must onboard environmental knowledge, which drives tactical decisions. Every CISO has had some involvement with SA. They use the information gained to influence decision-making, so security teams generate monthly reports, deliver status updates, and answer critical questions during executive meetings. As a CISO, the information serves as their knowledge base and decision-making authority. Imagine a corporation where the CISO is unknowledgeable about security practices. When emergent situations arise, the ability to make tactical decisions may fail or cause serious risks. Alternatively, when the CISO understands their technology landscape and has accurate information, they are positioned to improve security.

Beyond the CISO's involvement, additional thinking is required, such as cultural knowledge and cyber defense. Many computer network defense (CND) teams utilize operations and corporate security practices as a norm. Daily they approach situations that require "spot-on" responses. When incorporating SA, their defensive tactics and cultural knowledge separates success and failure. Imagine a situation where a culture of cyber thinking performed subpar during CND operations. The entire defensive approach and incident response activities would either fail or induce additional risk. Just a tiny security gap in cultural knowledge can harm CND operations.

Increasing cultural knowledge is a gradual process. It requires organizations to think of offensive and defensive security practices and innovative ideas to develop their mental database. Offensive security demonstrates internal situations and how the organization should respond—such as insider attacks. Internal teams place a significant burden on security, as they control conditions such as visibility and insider threat identification and activity. These are as simple as an employee executing unauthorized changes to system defenses or configurations. Defensive security starts at the "security gate"—commonly labeled as the perimeter protection boundary or external security zone. Typical operations require firewall settings and traffic blockage to occur, so with a SA mindset, one would be thinking about what should have access or restrictions.

Grasping cultural knowledge requires understanding the CND architect and its technologies. The architect represents a full scope of all the services and ongoing operations to include time-based information. For instance, organizations operate their IT systems based on specific IP addresses and traffic patterns. When integrating the CND theory and SA, a security analyst can identify different IP address scopes and respond when their practices change. The protection steps and reporting characteristics provide situational response and allow the organization to execute defensive thinking due to active engagement and thinking security. It's important to note that active engagement is a current state of security. Alternatively, the reactive condition is after the fact or incident occurrence. A functional SA program operates and gains success when active protection is practiced.

Active engagement and SA functions well together. When using SA, technical cultures and organizations can better predict, detect, and respond to cyber intrusions and system failures. Each employee who defends enterprise assets or data repositories must have some formal security knowledge or experience gained through constant observation or repetitive tasks. As active engagement develops, so does the environmental expertise. Integrating the methodology into a culture norm occurs by default or through job roles and responsibilities.

Displacing employees in various roles promotes cultural learning and also develops their baseline knowledge and mental database. The entire protection schemes, defensive strategy, and technical culture provide various learning disciplines. Assigning different responsibilities can create a robust mental database, actively involved security professionals, and cultural learning. If security personnel engaged in technology but not defensive knowledge, their SA would resemble a technical expert and not a defensive-minded person.

Role assignments can be described as a skillset development

program. Various CND-related tasks require defensive-oriented personnel that can react and responds to cyber-attacks. Their entire reactions are not overnight intelligence gathering or "on-demand" reactions. Instead, it's a combination of environmental skillsets that grow over time. Looking back toward the inclusive culture, we have reached the growth-mindset mentality. SA cannot survive as a static program; there must exist various learning activities that develop skillsets. Over time, technology and culture changes dictate the SA process, role assignments, and skillset requirements. The program fabric requires ongoing analysis, development, and involvement, which matures SA and its security functions.

When the COVID-19 pandemic surfaced, unethical hackers seized the opportunity by creating scams and malicious email attachments. To add, cyber actors released fraudulent information concerning charities through email. The emails contained malicious extensions that were executed through hyperlinks. As with any malicious threat, it's a constant learning process that challenges our situational response— it's a decision made despite specific situational factors. For instance, Security Analyst A thinks and weighs different factors when making a security decision. There may be many security lanes to involve or travel—so considering risk factors alleviate induced risks. This was increasingly important during COVID-19 since the nation experienced a medical pandemic, so generating a technology crisis would harm the medical progress and risk management initiatives.

The cyberattack response procedures for COVID-19 were a new and emerging threat. It required a growth mindset in understanding how the technology culture developed COVID-19 response procedures. The outcome provided an immersive environment where skillsets developed and SA expanded, all of which required active engagements. The term active engagement brings upon ideas concerning the broad functionalities and cybersecurity labor categories. Security engineers, software developers, network security, database administrators, and project managers co-exist in a typical environment. The security responsibilities and involvement can vary due to operational characteristics and assigned roles. Personnel that works in a SOC environment focus is response procedures and system intrusions. Groups that work toward vulnerability management and compliance hold similar but different engagement strategies. They reduce vulnerabilities through automated system scans or analyzing reports. In the SOC, the teams practice response procedures, react when intrusions occur, and require knowledge concerning vulnerable systems to determine attack profiles. The VM team analyzes incident data to determine whether attacks occurred via unpatched systems or applications. Now, what's the value? When the environmental knowledge aligns amongst different teams, their roles and responsibilities can drive a successful SA program. Each position operates in a collective space to address and drive security by combining their function into SA.

While maturing through the cybersecurity career field, I experienced first-hand knowledge concerning SA roles and responsibilities. On one project, we experienced a network outage that lasted seven hours. The impact prevented network communication between several DoD sites and using HBSS. Many of the end-users relaxed and ceased working since there was network communication and email was unavailable. The security team and IT support personnel responded differently when the network failed. Since every system was offline, we developed a "what-if" mindset involving cyber attacking and potential intrusion analysis. Although the system was offline, our SA performed root causes and system health checks. The IT support personnel operated in a nonoperative mentality, which means they ceased operations. Now, why was there a difference in approaching SA? It all relates to active involvement and role assignments. If the IT support personnel were security-driven, their entire approach would profile a defensive mentality and demonstrate a well-tuned and protection-focused team. When an incident or event occurs, thinking security provides a human firewall, which is defensive.

The traditional practice that the IT support team exhibited was customer relationships. Their primary role required support service and resolution vs. engaging CND operations. Asking the team about potential intrusions or CND operations would seem very foreign. When involving or cross-training different roles and responsibilities toward security, corporations become security-minded, resilient, and hacker smart, which matures their SA programs.

As an active security engager, SA applies through all availability periods or events. Despite network communication issues or system resources failing, the CND attitude remains. Another term that references this concept is constant visibility, which is an ongoing requirement. Organizations that embed the idea into the corporate security fabric are successful. Constant visibility is the end-state that promotes SA.

One area that focuses on constant visibility is the security operational center (SOC). Personnel that works within the SOC engage in the analysis of information feeds, network traffic, and log correlation. Each provides visible content and information concerning system traffic and changes. Typical events include software changes, web-based connections, resource allocations, or unauthorized access. The responder's role is to analyze the data packets and dashboards. The information may organize critical events through timestamps and severity. The information and dashboard graphical views are subject to change and update, which means constant visibility must exist. For instance, an event log may display a timestamp labeled 02:12:34 with authorized access has been granted. A hacker may also attempt to log on later and generate 02:23:36, which means they were denied entry. When SA and constant visibility deploys, the SOC personnel's role is to investigate the change. Through their analysis, it concludes that unauthorized access was attempted and failed. Now, imagine where constant visibility never existed. The log data may display various events that were never investigated, which could become a potential intrusion. Remember, hackers start low and aim high, so testing access is a low-level strategy. A high-level system is disrupting business operations.

SA survival depends on various defensive strategies. The impact and implementation of those strategies are by default, which means that as employees involve SA, they think defense. The term "fit right in" is where SA and the employee align because of embedded practices. When role assignments and job tasks require SA, the employee engages and responds with answers. It's somewhat an automated process considering you cannot engage SA without a defensive mindset. Organizations that involve SA in their security fabric will win. Having a winning attitude requires various tactics and deployment strategies. We have discussed skillset development, active engagement, and learning. There are additional strategies that involve tactical procedures through computer networks called cyber warfare. It projects the same similarities as military warfare, except the enemies are digitally connected.

To compete in the SA market, organizations must plan and respond through real-time methods. The learning curve and strategy require an organized process that teaches active response. Developing the best-in-class tools and analytical programs are great, but deploying a culture that absorbs the defensive mindset is more valuable. Through SA, employees can think defensively and position themselves as human firewalls. Although it's more mental than physical, the human firewall protects enterprise assets, and its integration is not an overnight task. The collaboration requires knowledge development and active engagement practices that operate as one entity, which are enablers that build focus and situational awareness. Once it occurs, technical and non-technical professionals can think defensively and advance SA. The methodology also requires security professionals to remain vigilant, which onboards their situational knowledge. Stay alert, and let's operate with focus!

CHAPTER SEVEN

MENTAL FOCUS AND ALERTNESS

The military has historically used combat warfare tactics to engage enemy targets. Through the entire process, battlegroups, brigades, and air squadrons gained success and learned valuable lessons, and their warfare plan involved situational awareness. The whole process required "thinking before the enemy" and making tactical decisions. Commanders would stress that engaging the enemy requires a well-disciplined and focused team. In their conversations, military warfare history and staying vigilant were key terms and talking points. Considering combat warfare may last weeks or months, the commanders had to drive mental alertness and awareness.

When military personnel enters boot camp, they are mentally cleansed and conditioned to thinks as an airman, marine, sailor, or soldier. Mental cleansing shapes their military mentality and removes distractions. These objectives are challenging, and they promote culture change for the servicemember. Survival is based upon their ability to contain detailed information, carry out orders, and stay alert.

When referencing military history, the engagement with SA onboard the USS *Spruance* surfaces. As stated in the earlier chapters, the transition to the USS *Spruance* demonstrated an inclusive culture. The division officer selecting me to stand the air-defense watch onboarded situational awareness. There were moments when enemy aircrafts intersected security zones, and the air-defense coordinator's responsibility was to track the aircraft. The watch lasted for six hours and rotated around the clock and provided a six-hour break between the next watch. Despite gaining minimal rest, I remained focused and alert. There were communication circuits relaying aircraft identification and flight status. The identification and tracking names were frequently repeated—so as a watch stander, you had to stay current, understand changing flight paths and aircraft origin, and continuously update your mental database. Referencing an earlier discussion, the period represented environmental knowledge. To accurately gather the information requires constant focus and alertness. I still remember the captain stating, "Chief, who are those aircraft flying from the left?" My response communicated flight path, origin, and aircraft type. Thankfully, the alertness and remaining focused provided the captain with confidence!

Mental alertness plays a pivotal cybersecurity role. Just as the military warfare onboards tactics and engage with focus, the technology industry onboards the same through cyber warfare. Every SOC watch officer or lead security analyst relies upon experienced-based knowledge to strategically guide threat response, incident handling, or intelligence gathering. As each develops targeted knowledge and relevant facts concerning their security environment, they successfully make informed decisions. These decisions are organically grown by staying focused and alert and building a mental database that involves situational response.

The COVID-19 pandemic created new challenges and response activities for the government. Many workers mobilized and worked from home locations, and businesses shifted their cybersecurity engagement to think COVID-19. Although the move was warranted, the focus on cybersecurity became a COVID-19 initiative. The Department of Homeland Security (DHS) issued several essential warnings and guidance to remain vigilant and mindful of cybercriminals soliciting financial requests. To combat the challenges, the DHS Office of Operations Coordination (OPS) created a crisis action team to support COVID-19 situational awareness. The outcome brings to attention: *did the DHS reshape America's focus and remove distractions?*

Distractions can disrupt alertness and focus. Many IT professionals perform multiple taskings that require on-demand resolutions or ad-hoc tasks. Coupled with high-tempo environments, their entire workload can become complicated and very disorganized. As a solution strategy, they may shift their focus or "drop the ball." In the IT market, each person has dropped the ball once. There is never room for excuses despite the operational challenges, so IT personnel must stay alert and protect enterprise assets. Also, distractions are constant problems that occur, and they can generate mental stress, fatigue, or lapse in communication. The end result affects whether cybersecurity focuses on what's most important—securing the enterprise.

SA cannot survive when distractions are present. Disrupting SA can shift priorities, invoke unnecessary risks, and disorganize operations. Through experienced-based knowledge or routine training, the disruptions can be remediated. Training provides the most excellent flexibility and process to drive alertness. For instance, a security analyst may have to clear threat logs and monitor intrusions, which they gain experience in handling multiple tasks and prioritizing security events. In their training, a scenario can target their attention toward managing multiple taskings and organizing priorities based upon criticality vs. ad-hoc responses. Another reference that describes the situation is various security lanes. The concept requires a security analyst to extract and insert experienced-based knowledge into groups or risk profiles. Using their experience, the analyst can manage multiple risks and objectives due to their group or risk-related function.

When removing the security lanes, an analyst may demonstrate a tunnel vision—a single security focus. The process works well for a sole objective, but additional or external security objectives can become risk-prone. The key objective is "balance." Operating through single or multiple security lanes is common, but dropping the ball is a bad practice. Through single or multiple tasks, distractions must be removed as it defeats security. We want security to stay on the winning edge! To mitigate the issue, the analyst must incorporate peripheral knowledge.

The relationship amongst security objectives requires analysts to think outside the box, which builds their external senses. When faced with a single security objective, it's evident that knowledge can become stagnant. Typical security teams are trained to combat threats, not multiple security events and risks. The training occurs because no analyst will ever receive a 100 percent playbook on approaching incidents. Through learned behaviors, engagements, and job rotational duties, an analyst can gain peripheral knowledge. The outcome presents the opportunity to defend the most sensitive technology domains. In modern history, every security team has been provided a playbook, which is a virtual practice. It's defined as being virtual because unwritten procedures and engagement strategies are hardcoded into IT security. The term "hardcoded" relates to an embedded and culturally defined process. Early in the career progression as a security analyst, one receives the most basic knowledge toward defending enterprise applications and systems. As one increases one's learning curve, the playbook becomes larger and holds relevant strategies concerning outside-the-box engagements. As the cycle continues, the information, data collected, and response procedures become an embedded practice.

Corporations can design a successful risk reduction and mitigation culture when deploying a Cybersecurity Mindset and peripheral thinking. The entire security domain and its threats become a "thinking cap." The collection of relevant facts and data feeds are analyzed and resulted as "what else could occur." At this point, the security engine operates at full throttle. Teams actively engage security objectives and think about who, what, why, and when to protect resources. The entire cybersecurity image becomes much more comprehensive, and their knowledge gap increases, which promotes the growth mindset.

In 2012, I approached a similar experience concerning peripheral knowledge. While working on a vulnerability management project, our team received a critical task from the enterprise manager. The environment had over 300K Java vulnerabilities that were deemed a Level 1 security risk. The environment had previously created a vulnerability management plan that outlined remediation practices. The plan provided guidance on patch deployment, remediation steps, and reporting standards for the enterprise; however, it neglected advice on remediating 300K Java vulnerabilities or had an existing playbook.

When referencing the Java project, the concepts for peripheral knowledge were relevant. It was due to our team requiring external knowledge and possessing basic remediation guidance. Here is where thinking outside the box played a pivotal role. In the environment, Windows desktops and servers had various operating systems installed and multiple Java vulnerabilities. The remediation plan stated, "Remove all deficiencies from Java with the current patch." Let's think for a second: the current patch should support all operating systems. Of course, the remediation plan never stated which operating systems; this is where peripheral knowledge surfaces. To be successful, our team had to know that every operating system required a different Java patch. Staying mentally alert and knowing the patching environment made our jobs easier.

Peripheral knowledge cannot operate alone when deploying mental alertness. One of the enablers that promote the methodology is maintaining awareness—it demonstrates a person has attentiveness and a watchful mentality. Typically, any change in system state and events would require mental alertness and attention. Some typical examples are timestamps that change regularly or external resources that affect the timestamps. A security practitioner that thinks outside the box would have the necessary skillsets to understand that previous events and system states change. They would further use the skillset to analyze security incidents and accurately respond to emergent or incidental changes.

Outside the determination of events and changes, awareness must also include resource dependencies. There are outcomes where security events require external resources to satisfy a task. The entire process may require indicators to only function when firewall changes occur or a system transitions online. Before the event, the timestamp may remain stagnant and cause some reactions: "Why is the system not changing?" Using awareness helps to resolve the concern. An analyst who stays aware of the current environment understands that the indicators are externally driven and mentally remains alert. What does this mean? In simple terms, they can accurately execute decisions and implement safeguards, mitigations, or remediation plans.

Verizon Wireless conducted a data breach investigation for FY2019. The report provided data breach findings for victims, initiators, tactics, and commonalities. The report demonstrated that areas for social attacks, malware, hacking, and motivational strategies largely contributed to breaches. It provided a common finding that showed 56 percent of breaches required months or longer to discover. When examining the commonality, the root cause highlights that the security team's operational knowledge failed. If situational awareness operated, the indicators and resource dependencies would succeed—a resource dependency requires another resource to operate. As a security steward, you have tools and information to identify data usage, system access, or data transactions. The information does provide when, where, and how the data was accessed and extracted. So, the story goes, how could the breach require longer to discover? Many drivers may cause the breaches to go unnoticed, and understanding resource dependencies stand out as drivers. Simply meaning the operational environment various resources necessary to identify the breaches, and each has a dependent. The lapse of knowing what, when, where, and how the resources dependently operated was a failure.

When breaches occur, system logs provide access control and timestamp information, and IT professionals can use the information to verify data records and system checks. When performing the security checks, the indications are extracted from their mental database. The information is further used to conceptualize whether the system indications are normal or abnormal—this is where the awareness surface. Relating to the Verizon report, it's apparent that the breach discovery period required months due to resource dependencies being overlooked or system checks poorly performed. I often relate to "what's on your network" as a serious concern. Throughout the non-discovery period, security practitioners examined logs and performed security checks—so why was the breach overlooked? Lack of awareness is one reasonable answer. Another concern could be a lapsed security environment and its risks. Its outcome affects situational awareness and demotes environmental knowledge. As with the Verizon report, there were probably indicators available, but through knowledge gaps and minimal security engagements, the breaches survived long. Another possibility may have been a lapse or misunderstanding concerning continuous engagement.

The term continuous engagement is an ongoing involvement that drives situational knowledge and develops security focus. When referencing the inclusive culture, constant attention is a cultural practice that guides the environmental experience. If the security practitioner stays engaged, their experienced-based knowledge expands, which relates to "Branding a Training and Awareness Profile." The complete process starts early in the security development stages and navigates through situational awareness. The culture identity brands situational awareness, so that continuous engagement and alertness can provide identical strategies.

One of the sub-categories relating to continuous engagement is security visibility. The term refers to sustaining a 360-degree awareness into an enterprise security architect or window. The enterprise defines all the components, programs, technologies, or policies. At the same time, the window is more streamlined at one particular system, process, or event. When deploying the 360-degree visibility platforms, the focus is to engage in enterprise awareness and sustain security readiness.

When relating to the Verizon report, it's apparent that security visibility failed. Each enterprise awareness program incorporates security visibility into its architect or window. SOC personnel utilize tools that provide graphical views, data representation, and security logs that offer real-time indicators. When engagements occur across visible platforms, organizations can stay focused on securing their enterprise assets and remediate security deficiencies. Security visibility can be described as the "scope" that remains focused and drives alertness. Part of the success depends upon understanding the standard and abnormal visible content and threat indicators.

In 2005, the USS *Spruance* was in its final deployment stage. The ship's mission required extensive sea time detecting aircraft and protecting international waterways. The combat engagement team was focused and ready to activate weapons defense when needed. Every combat watch stander learned various situational awareness and threat visibility disciplines throughout the ship's training cycle. Although it was shipboard-related, the concept applies to cybersecurity. The ship's window provided visible threats due to the prior training, continuous learning, and the crew experience–based knowledge. For sailors, this is what made the sea life secure and functional. Our shipboard motto: we are ready to fight!

One night, the *Spruance* "warfighting" mentality was activated. While sailing in the Persian Gulf, the ship's combat team received taskings to track and identify three hostile aircraft. Since the aircraft was flying below the hostile-altitude range, the team positioned the missile defense system to engage the aircraft. None of the watchstanders were told how to respond. Throughout the repetitive training, the combat team conceptualized how visible threats were processed and engaged. Since there was a continuous engagement for combat defense, the ship stayed focused and alert. Everyone knew when, how, and what responses were appropriate concerning the given situation. In the cyber community, the same applies, and security visibility drives proactive engagements.

Proactive security strives to think ahead and focus on what matters most: defending the enterprise. Every security analyst works to sustain visibility and environmental awareness, so assets and information can securely operate. The action requires attention and cultural norms, which means complacency is obsolete. In a SOC environment, it defeats security when complacency is activated. The required alertness and response may be overlooked and develop serious harm such as hacker intrusions and system unavailability.

In the cybersecurity community, having a complacent mindset shifts priorities to self-satisfaction and unawareness. The security team focuses on what they foresee as important and divulges attention from their security responsibilities. The term "work harder" is non-existent since the focus and situational awareness is obsolete. Part of the issue relies upon feeling protected after assessments or not understanding protection needs. Having satisfied protection today does not guarantee future enterprise protection. Shifting into the Cybersecurity Mindset onboards situational awareness and removes a fixed mindset. All personnel involved know their responsibility and security actions required to secure enterprise resources, assets, or information.

Satisfied protection stems from assessments and risk reporting. It's great to have confidence and trust that an enterprise security posture passed a recent evaluation; however, it does not replace responsibilities. Organizations must always utilize security visibility, peripheral knowledge, continuous engagement, or active security practices to sustain awareness and remediate assumptions.

When performing security assessments, organizations can use various metrics to determine systems' cyber hygiene and trustworthiness. Some organizations use percentages or scores, while others use weighted factors and ratios. Despite what is used, the idea that security protection removes risks is dangerous, which is an excellent example of complacency. The Verizon data breach had similar challenges where information and threat levels complied, but deeper security concerns existed. Their enterprise may have used security benchmarks that resulted in a successful compliance score or satisfied risk level. The knowledge of whether complacency occurred was probably unknown due to false awareness and minimal involvement or previous assumptions that "we are safe." As previously discussed, organizations' situational awareness is a vigilant practice where complacency is remediated through continual involvement and peripheral knowledge. These are the enablers that foster experience-based knowledge and influence security personnel to understand satisfied protection.

I remember conducting a security control assessment for a federal agency. The outcome of the evaluation yielded 95 percent, and every security steward relaxed and became complacent. In their mind, a 5 percent failure was not harmful. Well, a 5 percent factor can carry ransomware or other malicious codes that ultimately bring down networks. When situational awareness thrives, the 5 percent factor is continuously examined and assessed for vulnerabilities and results in staying alert!

A pass or failing score should be stressed over metrics. When the cyber mindset shifts, the concern addresses whether we are protected and whether situational awareness aligns with our strategy. The strategy may involve different awareness requirements and indicators that state, "We want to pass," and operate in a 360-degree visibility platform. At this point, cybersecurity thinking escalates and drives high-end remediation and protection practices. These statements do not entirely remove metrics but offer a detailed procedure to address complacency and sustain awareness. Cybersecurity thinking is mental and visual, so showing a percentage factor can distract the overall intention to pass the assessments. Metrics should be used as indicators and not the end state or primary driver. All too often, the numbers do not include the complete platform. Technology professionals gain 360-degree visibility by extracting the entire picture, which is beyond the 95 percent score.

A complete situational picture is a learned process that grows with technology teams. As the team develops and engages security, their experienced-based knowledge and mental database become important. Both provide a storage location to extract information and stay focused. The entire environment then shifts into a deeper mindset where awareness escalates, and decisions are trusted, but not overnight. It takes a technology team time to gain knowledge through practical and active involvement. Once the goal is reached, they can achieve trust and build their intelligence and automated response capabilities. Imagine a situation where experts predict, detect, and respond to security events through cyber sensory—early insight, predictions, and information concerning future security events. The entire skillset would defend the most critical systems and outsmart hackers. Let's activate our cyber senses!

CHAPTER EIGHT

TRUST YOUR CYBER SENSES

The human senses consist of touching, sighting, hearing, and smell. Each provides the ability to understand out-of-body experiences and environmental surroundings—such as entering a home and noticing changes. If a homeowner practices visual interpretation, they can distinguish objects in abnormal positions. Without a visual sensory, the homeowner would overlook the item and relate abnormal as normal, which is an awareness failure. The same path is often followed within technology, and the failure arises from inaccurate information or inactive engagements. If involvement and awareness operate, normal surroundings are sensed!

Senses are vital, and they allow humans to distinguish what, when, and how to react to events and situations. Visibility provides a physical view of conditions that are unsafe or require emergent reactions. The hearing determines how noise level builds awareness and attitudes. When humans hear unfamiliar noises, they react or respond due to situational changes. The same concept applies to a mechanic—they have an exclusive insight into engine noise. When "Bessie" starts to make an unusual sound, a mechanic may state, "She's not happy."

Touching provides physical sense into the surrounding objects, while the smell is a single category. Despite how the senses operate, they provide the gateway to engaging situations and events and using the mental database to determine normal or abnormal conditions.

The technology industry is no stranger to using senses. As a problem-solver, technicians gain an understanding of problems and their resolutions. When receiving an initial report, a technician initiates their thoughts and conceptualizes the root cause. It's fair to determine the root cause since they have background information and preexisting knowledge for the problem state. It sounds like experienced-based knowledge interrelates across the board!

The cybersecurity industry uses senses to engage and make risk determinations. Through environmental knowledge and active engagement strategies, security teams can determine and resolve problem states. When SOC personnel integrate visibility into their working relationships, they can distinguish whether events are normal or abnormal, which stems from their mental senses. The technology industry relies upon mental senses because it's an extension of the mental database. All the learned information, engagement activities, legacy problems, and security involvement practices are held in the mental database. Once an event occurs, the cyber senses activate, and they apply their learned knowledge as problem solvers. As stated before, the senses serve as learned behaviors and reactions that technology professionals engage. To deploy a cyber sense mentality requires a detailed understanding of what, when, and how situations initiate and their response level.

The Cybersecurity Mindset drives the actions to safeguard and defend critical systems and applications. It uses cyber sense to provide actionable intelligence. The alternative meaning for the strategy states that SOC personnel can respond quickly and efficiently without any distractions. Actionable intelligence offers an instant analysis of what is essential. Security personnel analyzes multiple data feeds and information, but only a fraction of the report carries value. Big Data, SIEMs, and various data feeds are used to collect security data and drive risk reduction. Analysts must decipher what feeds are actionable and how the data is analyzed. Through using cyber senses, the analyst can quickly label and identify valuable data. The end state provides intelligence concerning situational awareness.

Actionable intelligence operates through various mediums. Through each medium, there exist data feeds and information concerning security events. Various applications such as Splunk, Nessus, IBM Big Fix, or LogRhythm provide the data. The analyst must decipher the data feeds. These applications are automated tools, but the cyber sense is a mental approach that utilizes technology, and its informed decisions require accurate and reliable data. The statement does not define automated tools as useless. However, it does stipulate two mediatory levels that drive cyber sense—security automation and mental awareness. As previously discussed, mental awareness sharpens the response time and helps decipher data feeds. Within any situational awareness cycle, data that is accurate provides insight and quicker response time. The engagement into cyber senses and making informed decisions is not overnight; they rely upon environmental knowledge, peripheral vision, and staying security-focused.

As a vulnerability analyst, I worked in various analytical roles deciphering data feeds and reports. Many of them had aging history beyond ninety days. In a security sense, the most current data was considered actionable. Outdated data existed due to decommissioned systems or scan errors. There were moments when clients would ask, "Why does this report show old data?" and as a response, our team would resort to the automated tools. The entire cycle was a developmental action where we learned the environment and the systems' operational characteristics. Our developmental process started with minimal understanding of the situation. To achieve better experience, the team held meetings with subject matter experts and learned which systems or applications were critical and their current state. As we gathered the information, our mental database and environmental knowledge became much more powerful. Later, the information leverages our cyber senses. We knew particular systems that would fail and how to deliver informed decisions. Thinking outside the box, peripheral knowledge, security focus, and actionable insight became one entity. Our cyber senses provided valuable insight and accurate information. We rarely researched what constituted aging vs. non-aging data or "how to respond." Each vulnerability report was cyber sensed!

Data is a core enabler and drives cyber sense, and the entire lifecycle consists of creation, use, storage, share, archive, and destroy. Data creation is done through security events or use. Once the data is used, it's stored in a virtual or physical location. For the cybersecurity community, events and threat feeds are stored within databases. Sharing occurs by system-to-system, security groups, or individuals providing data access.

The cybersecurity data lifecycle starts with raw data. When events display timestamps and threat changes, the contents are described as a Tier I Data format, which means the data is raw or natural. Once the data is analyzed, it's grouped based on severity, impact, source or destination address, and threat level. The outcome then results in Tier II information—defined as processed data. Typically, applications such as Splunk will provide Tier II status. Splunk is a data aggregation tool that many IT security teams use to massage big data into a valid format. How the information is used, described, valued, or labeled defines Tier III—actionable intelligence.

When dissecting the data lifecycle, evidence proves actionable (value) intelligence (processed data) and has a roadmap into the cyber senses. Whether it's upstream or downstream data, the lifecycle must be understood. Environments rely on accurate and actionable data. When deploying cyber sensors, understanding the information origin, advanced process steps, and reporting requirements affects decision making. Using senses that rely upon assumptions can weaken security defenses. Knowing how the data affects the enterprise and severity carries the most weight and helps determine the security impact.

Another critical point concerning actionable intelligence is "checking the checker." A modern term is quality assurance, but a more formative name is data quality management (DQM). "Bad data inserted equals bad data out" is not a new term; instead, it's an overlooked concept promoting risks and loss. Thanks to cyber sense, checking the checker can provide a second-level decision or feedback. The idea is well suited when reporting information to executive leadership. When leadership validates the information, DQM and cyber senses become one entity. They gain an understanding when reports are inaccurate, and they propose questions, which is true to say, "If something seems wrong, our cyber senses must assume authority." Even a CISO must stay engaged, alert, and fine-tune their cyber senses and be "checked."

Checking the CISO is just one puzzle, so adding different pieces such as a questioning attitude can also become beneficial. A questionable attitude is a great skillset and pathway to cyber senses. Sometimes, the data or events collected may seem undervalued or less critical, but the decision becomes successful through senses and experienced-based knowledge. Every CISO has dealt with similar situations and relied upon internal and external information to make decisions. Through internal meetings and different reports, they gain sense into approaching tasks and guiding their cybersecurity vision. Over time, the story becomes experienced-based knowledge and activates when situation awareness surfaces. Their expertise or mental database relies heavily upon downstream resources and dedicated security teams. The downstream resources are lower management, application, advisory boards, and various technology platforms. When all else fails, the CISO reports to external support.

External support, such as briefings, focus groups, and webinars, provides CISOs with an arsenal of data. How the CISO uses the data determines whether it becomes information and actionable intelligence. These webinars discuss technology trends and various security changes for the industry. Although the CISO does not operate at lower-level situational awareness, they still engage SA strategically. When different reports surface concerning upward trends in vulnerabilities, a CISO's cyber senses provide relatable information and responsible action. The outcome may require the CISO to change the environment threat or risk level or refocus the vulnerability remediation and priorities.

Shifting priorities is a learned experience and requires a detailed understanding and a 360-degree involvement. As a CISO, the main contributor to decision-making is actionable intelligence and accurate data. The information is usually shared and communicated through various channels such as executive meetings or project-related sessions where teams distribute advisory information. Defining accurate data is where the cyber senses become essential. If the CISO's 360-degree involvement is active, they can identify data errors or inaccurate reports and collect actionable intelligence, which heavily depends on thinking "defense."

Executive leadership involvement in SA goes beyond the boardroom and meetings. Although they are not heavily involved with hands-on security, their position still affects their defensive strategy. The boardrooms typically hold discussions based on the number of incidents, performance metrics, or key risk indicators (KRIs). The information develops the executive leadership's mental database and cyber senses. Through regular reports and advisory notices, they can react and drive organizational changes that meet the mission requirements. Onboarding performance metrics provide leadership with actionable data concerning future attack probabilities and areas not meeting service level agreements (SLAs). The entire process operates the same as a lower-level manager involvement, except the executive leadership is more strategic. Their role is to decipher intelligence data and use the knowledge gained in shaping the organization's cybersecurity program. The leadership team intelligence is by far their best tool—as it provides experience in deciphering what's accurate. For example, a CISO may receive an incident report that shows upward trending. The data may state February 234 and March 434 database incidents. The CISO priority is, "Why are we up trending?" which is their cyber senses. When using their mental database, CISOs can extract previous knowledge and decipher the report accuracy. Typical knowledge such as whether databases are online, previous reports, or mitigated incidents surface. Let's imagine that the report shows Database_236 is the affected system, and it has been offline since February 28th. The CISO may determine the report is inaccurate due to their knowledge. If the CISO accepted the report as an action item, the entire enterprise would waste labor, time, and funds to remediate a non-existing security issue. The result: the performance metrics would create a failing strategy. A failing strategy falls within strategic planning, which is just one cybersecurity pillar.

SA occurs across three primary pillars, which are tactical, operational, and strategic. These pillars mimic warfare levels, and each particular pillar provides an advanced form where cyber senses are deployed. The CISO is the most advanced pillar (Tier I), which is strategic; tactical is one step lower (Tier II), while operational is a ground-based pillar (Tier III). Despite which pillar is being deployed, organizations' cyber senses play a vital role in defending systems and engaging security through intelligence.

The tactical approach disguises movements and deploys counterattack scenarios—such as blocking network traffic from an unauthorized source or, as previously discussed, responding to network attacks. In the modern technology scope, tactical engagements are a Tier II process. The pillar resembles the attack responses and security tools such as Splunk, ArcSight, or Tenable Nessus to discover vulnerable applications, events, or threat indicators. When levering a Tier II approach, the cyber senses require visual data and log information to detect, protect, and defend assets, services, or data. The response team leans heavily upon hands-on skillsets to decipher events, threats, and intelligence. The perimeter defense skillsets are vital in Tier II approaches. The response time requires instant involvement, which means your cyber senses are in activation mode.

The Tier I operational strategy supports the information system needs. Under this particular strategy, a corporation works heavily to sustain systems and application availability. As a security manager, one would drive their team to sustain availability based on cyber senses and knowing the system risk state and requirements. The entity or agency may have different scenarios and risk relationships, so as a security steward, the environmental knowledge becomes a significant factor—here is where you learn the business units. Incorporating the learning mode allows the technical team units to onboard each security involvement and define the business requirements. Typical service level agreements (SLAs) provide the elements and help to identify when security needs "calibration." As a security steward, the cyber senses would require the use of system uptime and downtime. If systems are offline for an extended period or beyond the SLA requirements, there should exist a mental alert and cyber sense capability that states the SLA "needs attention."

In 2012, I transitioned job assignments and started to support extensive vulnerability management (VM) platforms. The taskings from the previous roles supported risk and security assessments and developing the documentation or artifacts. The VM role required constant analysis and reporting of critical security issues with laptops, servers, and third-party applications. Monthly, the team would execute vulnerability scans using Tenable Nessus—Network Security Scanner. Nessus provided data concerning system configurations and vendor-recommended patches to close vulnerability findings.

Using cyber senses was one of the most critical practices for the VM role. The task engagement operated at all three pillars: Tier I, Tier II, and Tier III. The Tier I work required analyzing the reports to determine false positives. Our team utilized active involvement and engagement principles to assess the false positives. Each report contained missing IP addresses, NETBIOS, or DNS names. To add, we knew when Microsoft Patch Tuesday occurred—so when patch deployments were scheduled, our team could deploy our cyber senses and identify patch requirements that affected the environment.

The Tier I function provided an opportunity to understand the company's availability requirements. Different business lines operated on various requirements and technology needs. For instance, the application service units required server uptime to ensure critical applications operated on a 24x7 schedule. The patch deployment schedule coordinated the time needed and business line approval. We also had involvement and operational requirements for database systems. The database environment performed daily, nightly, and ad-hoc backups—so deploying upgrades and patches would be intrusive. Typical scheduling and upgrades were coordinated based on the different SLA requirements and operational commitments.

When invoking the strategic tasks, our team provided metrics that guided the enterprise in making informed decisions. Monthly reports were provided that indicated upward or downward trending for the vulnerability count. The data served as a source for the executive leadership to determine security gaps and areas requiring further attention. The reports contained data representing vulnerability aging summary, servers or laptops count, or critical vulnerabilities that negatively affected SLAs or business operations. The executive leadership further used the information to plan infrastructure upgrades and remediate critical advisories or CVEs.

The cyber sense methodology has provided a new design in addressing security. In 2019, the federal government issued a mandated policy to modernize its IT infrastructure and operations. The modernization plan, commonly referred to as the digital modernization goal, was to onboard newer technologies, operate the security landscape as a lifecycle, and design the federal technology space as an adaptable framework.

When referencing the cyber sense methodology, the entire modernization process served as a strategic vision for the government. They onboarded adaptation strategies due to security gaps and legacy management strategies. During the phase, some executive manager's cyber senses provided actionable data concerning legacy technologies since IoT, Azure, AWS, Cloud. So to keep abreast, the government's future would require technology alignment, which reduces costs and cybersecurity risks.

Deploying cyber senses occurs every day within the technology space. It supports the SA framework by helping security professions identify a "quick response" and an engagement approach that defines "defense." Combining the cyber-sense mentality with various business or technical processes requires learning. Trying to achieve overnight proficiency in cyber-sense deployment is far challenging than one could imagine. It requires a masterful skillset and a spectrum of knowledge. One of the critical areas that support the methodology and the full scope of SA is communication. In a more defined term, communication addresses the distribution of actionable data, information, or intelligence. The entire decision-making process utilizes a communication chain that integrates who, what, when, and where SA is deployed, driven through information sharing.

CHAPTER NINE

INFORMATION SHARING FOR SITUATIONAL AWARENESS

The technology culture has survived many challenges and performed well due to the knowledge gained and collaboration initiatives. Both present the flexibility to learn and gather actionable data, defend the most critical systems, and lower risks. Platforms such as United States Computer Emergency Readiness Team (US-CERT), Open-Source Intelligence (OSINT), or DoD Information Operations Condition (INFOCON) leverages a path for early warnings and communication pipelines that improves situational awareness and threat response. Every IT organization receives or distributes the advisories that communicate new or advanced threats. Their security professionals are masterful in deciphering and integrating response or mitigating the advisories. During the process, they refer to the collected information as intelligence since it's actionable. Typical information includes reference ID, vulnerability description, affected systems, threat level, and solution. The teams receive the advisories via email, and middle management provides regular updates or status reports. As a hands-on security professional, one must understand the Tier I engagement and information system personnel think "availability." At Tier II level, SOC teams must set specific defensive measures. The Tier III involvement approaches the advisories based on the organization's mission: can we support the customer? In the end, all three pillars require information sharing and actionable data to remediate or enact upon security advisories, and middle management has to openly share the data, which is a challenge within certain corporations.

The building blocks of understanding information sharing start early for a cybersecurity career. As a junior security analyst, I still have memories of examining threat logs and informational reports. Our team used various communication channels to share information for contractors, military personnel, and government employees. Over the years, some of the sharing processes changed, but for the better. One avenue that has not changed is the message system. While standing watches, there were many intelligence briefings and reports distributed to cybersecurity teams. The information always provided background information and action steps to remediate vulnerabilities and monitor threat actors. As skills developed, sharing cybersecurity information became the key to incident reporting and analysis. The timespan between an actual incident and the remediation period required quick action and data collection skillsets. The success embodied one point: early information sharing was vital. When citing the junior security analyst, the role was a breakthrough and an avenue in learning how cybersecurity and situational awareness shared mutual dependence. Each participant shares a common interest in data protection and establishes a trusted relationship. The US-CERT plays a pivotal role in guiding the US cybersecurity infrastructure and information sharing.

The US-CERT is a core enabler and vital contributor to initiatives concerning situational awareness. The informational sharing has saved many lives and "outsmarted" hackers when developing the country's infrastructure. At the forefront, the US-CERT Cyber Information Sharing and Collaboration Program (CISCP) supports sharing incident data, cyber threats, and situational reports. The program offers various services that navigate the situational awareness process and concerns. Every organization has information sharing and data dissemination dependencies. The US-CERT has developed milestones and sharing initiatives that guide cyber teams to stay closely connected and develop a "trusted relationship." Despite the effort, there still exist information-sharing gaps and individual concerns.

Having a partnering mentality provides the source for navigating information sharing. Partnership sharing is a methodology that provides early warnings and aligns situational awareness response cycles. Let's think for a second: hackers share information and stay closely connected, so can technology professionals embrace the same? Yes, they can model the same process, but a deeper understanding and shared security connection must exist. The Cybersecurity Mindset is a mutual process—so information sharing and inclusive culture are families. In Section I, the concept was discussed, and it described how cultural relationships operate in unison. As with sharing methodologies, organizations' situational awareness programs are culturally based, and it promotes partnerships. As a result, partnering companies and business lines communicate the same sharing mentality and cultural practice.

In 2018, NextGov posted a very informative article concerning information sharing between the Department of Homeland Security and non-federal entities. The report, titled "Only 6 Non-Federal Groups Share Cyber Threat Info with Homeland Security," highlighted that actionable cyber intelligence was not shared. The root cause was argued that corporate concern or individual preference to protect cybersecurity exposure was the fault. Although the solution sounds legitimate, culture sharing has its rewards. The transmitted information concerning organizational data can help predict situations and prematurely warn off attacks. One point to make: none of the organizations were directed through an executive order or governmental regulation to share their information, which means the sharing initiative was an incentive. The sharing initiative was supported by the Cybersecurity Act of 2015. The act established mechanisms for cybersecurity information sharing among private sector and federal government entities. When referencing the SA framework, the action prevents incident response failures. Also, it supports an aligned threat picture. When the DHS and commercial sectors share the same threat information and data views, each can gain early threat indications and deploy countertactics. In the context of the Cybersecurity Mindset, the act provides a direct gateway to building the most cohesive information sharing process and a simplistic approach: *why, what, when, and who* should view the data.

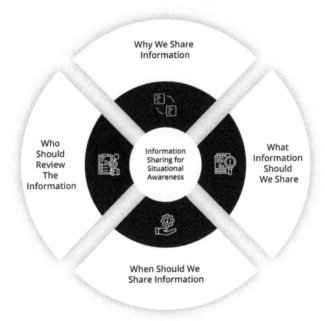

DIAGRAM-1: INFORMATION SHARING CYCLE

Diagram 1 provides an example of an information-sharing model. In the cybersecurity community, the why, what, when, and who methodology allows the analyst to deliver informed decisions. When they are challenged to approach incident response or security events, the model can connect decisions. No analyst will probably view the model due to emergency requirements, time constraints, and rapidly changing events. Still, it can serve as a preliminary learning and development tool that prepares SOC teams to think and respond quickly.

Integrating a sharing methodology can become challenging and requires trust. Many organizations are hesitant and protective of their security information, so the initial step is to structure internal programs that work, which will, at the minimum, provide external trust. The trusted relationship allows sharing partners to determine an organization's success. Owning a failed product can destroy information-sharing capabilities. That's why a structured process is required, such as the Information Sharing for Situational Awareness (IFSFSA) model.

The ISFSA model is a procedural-based approach that drives situational awareness. The ISFSA concept fails when ad-hoc information or processes are included. Attempting to manage a SA environment through unstructured practices and responses is risky. The ISFSA facilitates the Cybersecurity Mindset and engages how responses and approaches should operate. Diagram 1 provides a high-level view into a model that supports ISFSA and its functional processes. The model supports all Tier levels and addresses key areas that drive situational awareness across every security platform. The five elements are quick factors in determining an organization's mindset and weakness within its engagement strategy. The features compare to data visualization programs, which is a primary source for making informed decisions. Data visualization relates to what security stewards view and present, such as graphical views, logs, reports, and guides. The visual data provides information that defines who, what, why, when, and where they should present data. Let's initiate a deeper dive into dissecting the model and its strategy.

The first action, "why we share information," is an initial thought of external partners' data exchange. The concept can also be easily described as "why are we sharing information" information.

Organizations must first define the value or reason to share data. The answer relies upon policies, guidelines, and security incidents, all of which guide decisions. Most SOC personnel are not directly involved with policymaking, but they must adhere to policies. Today, someone in some security function is sharing data without thinking, "Why?" which is driven by business or situational norms. Understandably, SOC personnel are heavily pressed to make quick decisions, but sometimes it's smart to "just think logically." One quick reaction without knowledge is dangerous. Situational awareness thrives in the framework of experience—if you are unsure, ask someone.

Tier III requirements drive the "why" factor. From a high level, the governance team communicates the corporate vision and data-sharing strategy. In my previous experience, I have seen many data-sharing initiatives and advisories that direct security teams on "why" data is disseminated amongst partners and institutions. Then, there have been times when teams "shoot from the hip," which means without understanding "why," they pressed the button. Working as a team in understanding why actions occur is much more productive—it provides a structure for decision-making. In relationship to the inclusive culture, the same concept applies where teams operate in unison. When everyone understands the corporate vision, they onboard "why" sharing occurs, and they can use a "buy-in" mentality.

Aside from the Tier III approach, the Tier II strategy is more tactical and builds a Cybersecurity Mindset into why we are sharing data. Before initiating the incident, event, vulnerability, or threat condition, it provides direct interaction and the decision point. A simple answer could be: *It gives "Company A" information on an Operating System (OS) threat that affects Linux servers.* The approach is more technical and assesses the security event or incident in question. The "why" factor still applies, but it examines technology decisions compared to Tier III, which is more strategic. Tier II can be much more damaging when just "pushing the button"—as it's the stage where information is transmitted and the "what information should be shared" factor comes into existence.

The transmitted information has a direct purpose and benefit to making decisions. However, in the interim, a security analyst must always understand information flow based on the "what factor." Every organization realizes that sharing information is helpful and supports awareness. Tactical decisions require the soundest cybersecurity information rather than scattered details. Just imagine trying to understand unorganized and unrelatable information—the process would create significant delays in responding to events and emergencies.

In every situational awareness platform, cybersecurity information is analyzed and transmitted through concise channels. The sender holds the primary responsibility of ensuring the reports are relatable and valuable. The recipient receives the data and makes actionable decisions. Their decision to control what is shared is limited. So, what if the information was not applicable? The result would provide risky decisions during incidents or create a downstream effect where situational awareness becomes normal when it's abnormal—commonly referred to as pass-down knowledge. To remediate such actions, organizations can succeed by examining the "what factor" based on the situation or information classification scheme.

Every logging software program or threat logs provide repetitive information. Over time, security analysts can rapidly determine what is to be shared. During my early career as an HBSS security analyst, I experienced the "what factor." Our team routinely received threat feeds and log data for hosts and network traffic. We would examine the log data and decide on what actions were necessary through our analysis process. The severity category and security conditions drove the actions. For instance, the Navy would distribute a message that "malware" incidents were categorized as a critical priority, and the incident response guidelines handled Low, Medium, or High events. Typical guidelines operated through attack types and impact. When sharing the information, our team would reference the incident handling techniques to determine what information was shared. Attempting to share irrelevant information would draw confusion and prolong decision-making or cause data leaks.

The "what factor" parallels additional classification schemes such as Confidential, Secret, and Top Secret information. The entire sharing pipeline is guided through trusted relationships and security clearances. When situations warrant a response, military and government agencies think through cybersecurity protocols to determine what is shared. Based on the recipient clearance level, information is legally shared that matches their information classification status. The sender or recipient can either be a person, system, or application program with a trusted relationship. So, when analyzing what information should be shared, the sender must adhere to the need-to-know principle—only share what's required. The chain does not stop here, so knowing what to communicate leads to when to share information.

Managing time-based decisions is a skill that's acquired over time and requires experience with the operational environment. Information sharing becomes the "when factor" and helps contain attacks and rapidly communicate threat intelligence through every decision. Each threat management platform processes the same methodology, and SOC personnel is constantly engaging in time-based decisions. The success or failure factor is primarily guided through alertness and knowledge—critical drivers for situational awareness. The amount of information shared can be overwhelming, and the "what factor" shortens the list, as it provides the blueprint to decomposing events and threat information. In the Cybersecurity Mindset, an individual thought follows the sequence of "what to share" and then "when the information should be shared." In reality, the entire process is a skill that's gained through experience and environmental knowledge. It essentially reduces risks associated with information sharing and drives a successful situational awareness program.

The relationship between the "what factor" and "when factor" onboards information classification, severity, and clearance levels, but their position within the "when factor" drives time or event-based decisions. The time-based process is when system resources are at risk, availability is a concern, or threat intelligence information is critical. Its engagement, communication, and sharing strategy drive early warnings and cyberwarfare. The term "fighting the enemy" is a mindset and warfare tactic that strives through time-based decisions. A small gap within the information-sharing medium or lack of understanding of the "when factor" can affect the situational awareness process. Typically terms such as, "*I did not know,*" "*The information was incorrect,*" or "*When did you send the information?*" are common responses for a failing "when factor."

The time-based analysis is just one subset for the "when factor." The event-based analysis is another factor that's driven by criticality. When organizations utilize event-based decisions, their dependence follows security classification, clearance levels, or security incidents. The same concept applies to the previous discussion concerning vulnerability management tasks. The monthly reports contained critical findings that were rated as a high priority. The reporting guidelines mandated that critical or high vulnerabilities must be reported within twenty-four hours. Using the information-sharing process, the team created incident tickets and advisories that described the vulnerability. A Low or Medium event was analyzed and remediated within thirty days. Then the team had to decide on who should receive the data—it sounds like information classification provides an all-around foundation. A significant drawback that occurs when "who" fails is a security violation. Data may be presented or transmitted to an unauthorized source—so that's why background investigations and clearance levels are required.

It's straightforward that who should receive information impacts cleared personnel. As stated within the "what factor" section, every team or resource requires specific knowledge. The information can be categorized as host intrusions, predictive analysis, threat status, or advisory reports to contain specific attacks. Every impacted team categorizes its information based upon the current event or threat indicators. When information sharing surfaces, the end goal is to provide rapid and concise information and align the "who factor." Unlike the why, what, and when factors, the "who" is the end-user, system, or organization. The previous requirements, such as clearances, event descriptions, or severities, still apply within the "who factor." The result ensures that the appropriate resource receives information that is relatable and provides value.

Organizations' SOC teams are heavily engaged in situational awareness and the "who factor." There are third-party organizations, management teams, and internal groups that require readily available information. Some of the information sharing may be events or log data that display time-based access. For a SOC analyst, the Cybersecurity Mindset is one of the most reliable tools in deciphering the "who factor." It leverages the thinking capability, environmental knowledge, and information sharing principles as a single entity. The SOC analyst can use the methodology to ignite their sensors and transition the event into a communication and reporting medium. The results provide early warning, cybersecurity response, informed decision-making, and predictive analysis.

The information-sharing frameworks operate on principles outside the four factors, such as predictive analytics. It describes how previous data is used to determine potential events or incidents. In the financial industry, predictive analytics relates to forecasting as analysts predict a stock's future performance. In the technology space, analysts predict future threats, incidents, risks, or events, and these factors determine how information is analyzed and shared. In today's environment, organizations are moving toward digital modernization platforms. The technology automates data and provides predictive analysis. Its outcome can accelerate the time and labor required in a traditional situation awareness environment. However, it still needs some form of human intervention. Humans' requirements are to decipher the automated data as a false or true positive so the automated data can either be accepted or fine-tuned. The results do benefit the technology industry in correlating all the data feeds and various entry points. Imagine a SOC analyst trying to decipher five million records of log data. This is where predictive analytics provides the most value in streamlining information-sharing cycles.

As with many of the processes used for situational awareness, they give the most value when security gaps are remediated, and organizations can work through a sound situational awareness program. Reaching this level requires an in-depth analysis of their situation awareness model. The model includes all the previous information discussed and different scenarios that drive cyberwarfare. Defending organizations through situational awareness is where information sharing works the best, so let's learn about the assessment process and its contribution.

CHAPTER TEN

SITUATIONAL AWARENESS ASSESSMENT

E very security team has performed or contributed to performing assessments. The associated tasks require them to validate security controls and their effectiveness. The organization may deploy various tools and testing methods to determine the controls' operational and protection characteristics. Further analysis or sophisticated checks are performed to examine sampled applications, assets, or data for security deficiencies. Despite the selected test performed, the results and information gathered provide an in-depth analysis of the security environment. The results identify whether technical, operational, and management controls meet the security objectives. The evaluator may further conclude that the environment and security culture can operate with the minimum security, which is optimistic and can become misleading. Think about a tabletop exercise where a security responder adequately addresses an incident. Although the test is designed, it could influence the organization that designed testing equals normalcy, and the security objectives were adequately addressed. This can position risk occurrence as designed testing cannot represent every security situation. What do we know? SA is about normal and abnormal events. When companies establish the technology culture, active security practices, and the defensive mindset, they also create norms and abnorms. The confusion is evident during evaluations and exercises where designed testing is labeled as the standard.

Normal practices can be stagnant and describe an organization's comfort zone. Here is where security is labeled secure without any justification. If the practices remain and organizations presumably think their defense posture is the "Best in the Class," they have positioned themselves as a "hacker's meal." SA requires constant evaluations through normal and abnormal designed tests. A typical testing scenario uses the corporation's predetermined controls and security objectives. The abnormal testing is detailed outside-the-box assessments that evaluate responders' engagement. These specific tests provide room for improvement and early preparation into abnormal incidents.

DIAGRAM-2: SITUATIONAL AWARENESS
ASSESSMENT FRAMEWORK (SAAF)

For clarification, abnormal incident training is not intrusive. The word abnormal can become misleading, but for SA, it provides a more in-depth analysis of undefined scenarios and events. A great example is the use of malware attacks and their countermeasures. During the scenario, the team may use tools for counterattack exercises. The team may not mirror real-world attacks in an abnormal mode, so they inject outside-the-box scenarios during the testing phase. The outcome provides an opportunity to learn the corporation's security fabric. The follow-on requirements identify growth potential and build the culture situational awareness program and its assessment cycle.

The assessment cycle is described as a point where assessments are continuously used to evaluate situational awareness programs. There is no specific method categorized as correct. However, the assessment procedure must cover the security objectives and thoroughly assess a program. The cyclic process provides maturity and describes program readiness. **Diagram 2** provides an assessment model and system development process for accessing a situational awareness program. The model, called the Situational Awareness Assessment Framework (SAAF), follows the System Development Lifecycle and typical assessment procedures. Although strategies such as Analyze, Design, Develop, Implement, and Evaluate (ADDIE) are not mentioned, the concept is embedded within the SAAF. Each stage delivers typical assessment requirements and objectives and also determines whether the program is functioning. The phases are no different than risk management frameworks—they both provide a starting and ending evaluation process. The SAAF first stage requires an evaluator to learn the environment.

The assessment starts by conducting a preliminary review. Typical methods, such as organizational policies, SOC management, responding techniques, and time-based decisions, are examined. The evaluator would want to visualize and understand how SA operates within the culture and the buy-in mentality. The reporting guidelines and previous data are reviewed to determine the organization's reporting process, and security personnel may be tasked to provide live demonstrations. The overall objective is "familiarity"—it provides the blueprint for understanding the architect. Evaluators learn the "language" and "inclusive culture" profile. Typical notes are gathered that define daily tasks, communication pipelines, and administrative action. Another term that applies is, "How do they do business?" which supports a functional control assessment.

The National Institute of Standards and Technology (NIST) has published various publications to help facilitate the control assessment process. The publications provide a detailed analysis of selected security controls and their test cases. While performing several risk-assessment tasks, the NIST publications were used extensively. Several situations existed where different systems required a deep-dive analysis, and using the NIST publications made the task simple. Since the architect used web applications, databases, servers, desktops, and infrastructure devices, various control selection processes existed. Each had independent assessments and test cases that described the control features and adequate security measures.

The SAAF control selection process and its requirements mirror the NIST standards. Each control area must model the environment and provide security protection. The mirrored standard requires administrative, operational, and technical security controls to safeguard the program. For instance, decision-making controls would need the technical teams to use knowledge-based facts over ad-hoc information. The test case would examine intelligence reports, communication pipelines, or advisory notices to determine whether information-sharing utilized current and relevant facts.

Table 9 is a control framework for the SAAF, and it utilizes the most common controls involved with situational awareness. These functional areas may differ depending on the organization's structure, programs, or processes. For this explanation, three areas concerning administrative, operational, and technical security controls are tested.

Control ID	Control Name	Severity	Control Description
Administrative			
A.1	Corporate Buy-In Structure	High	The organization has implemented a comprehensive corporate policy and enforces situational awareness as a security enabler.
A.2	Knowledge Base Repository	Critical	A dedicated data repository exists where current and relevant data describe threat indicators and advisories.
A.3	Communication Plan	Medium	A communication plan exists that describes the situational awareness reporting, dissemination, and notification process.
Operational			
O.1	Information Sharing	Medium	Service-level agreements (SLAs), non-disclosure agreements (NDAs), and sharing partners' provisional rules exist, and operational teams share information responsibly.
O.2	Response Time Frame	Critical	The timeframe between the response time and closure for awareness events falls within its incident response (IR) guidelines.
O.3	Security Visibility	High	The operational team has 360-degree visibility into the corporate security architect.
Technical			
T.1	Tool Usage	Medium	Operational teams know and understand the cybersecurity tools deployed and their situational awareness support.

T.2	Notification System	Low	An electronic notification system exists and is tested monthly.
T.3	Alert Indicators	Critical	Alert indicators exist with flashing or "attention-based" colors that drive awareness of events.

TABLE 9. CONTROL SELECTION

The controls selected can change or require readjusting due to technology or operational needs. It's simple as an event response time varying from five to three minutes. The change would require the corporation to adjust the control description and testing scenarios. In the next section, "Access the Current Process," the assessment will follow the same concept and refine how the selected rules are tested and analyzed for gaps.

The integration of "Assessing the Current Process" is a formal evaluation and testing methodology to determine whether the control framework, organization's security objectives, protection needs, and execution aligns. Evaluators operate with an independence mindset— which means they serve as an investigator. They perform observation, tests, or administrative reviews. For instance, the investigator may choose to review T.2 for incident notification and monthly checks. The investigation outcome defines whether the monthly checks were complete and the notification process operates.

During the observation, the investigator observes the notification in real-time. The subject performs normal work routines and tasks, and the investigator documents the events. The investigator references various rules, regulations, response plans, and communication guidelines for compliance. They also observe the analyst transmitting or receiving notifications and validates its accuracy and content. Typical information investigated is time response and correct distribution. The investigator also reviews the communication plan to determine whether the organization receives the notification.

The testing strategy works along the same lines as the observation, except the planned and coordinated tasks. The investigator still reviews the policies and guidelines for compliance and validates the content and accuracy. The exercises are labeled as drills, and they are developed from the communication plan or previous notifications. Each particular exercise has an objective. In our example, the objective may read **Test Notification System.**

Each particular assessment strategy produces results that are either compliant or non-compliant. The information is used to identify process gaps and what specific controls need readjustments, corrective action, or removal. Over time, the program may change, and through the assessments, organizations can determine its weakness or security gaps.

One skillset that separates the Cybersecurity Mindset is understanding fundamental weaknesses, issues, or security gaps. Every assessment has its pros and cons. Some are more functional and realistic, while others are paper-driven. A more realistic framework involves the status of the current and ideal security state. The current state analyzes the security functions to better understand real-time effectiveness, while the desired state operates in normalcy or goals. The expectation is to deliver a functional cybersecurity posture. The paper-driven method does not incorporate real-time analysis since it drives administrative evaluations. A realistic model focuses on organizational, operational, and technical assessments, which are more functional and help identify security gaps.

The gap analysis requires businesses to complete a practical assessment, such as "Assessing the Current Process." The outcome provides realistic control status, and the reviewing, testing, or observations are included. A common term called "diagnosing cyber hygiene" is used to assess the control. The evaluators will use the reports and control outcomes to rate program risks. Once all the controls' hygiene information is compiled, the organization can then develop a gap analysis report.

Let's take a deeper dive into the gap analysis by choosing three controls. The objective is to evaluate cyber hygiene and its key findings. Table 10 provides a control list and their test results.

Control ID	Control Name	Evaluator	Test Results
Administrative			
A.3	Communication Plan	Semais.234	The plan did not include notification timelines or incident response teams that require situational awareness data.
Operational			
O.1	Information Sharing	Semais.234	Sharing between group A and the response team did not exist. The timeframe to share Level I incidents was not included within the information-sharing policy.
Technical			
T.2	Notification System	Semais.234	An electronic notification system exists and is tested monthly.

TABLE 10. TEST RESULTS

Our cyber mindset has two focus areas for the gap analysis—the current and ideal state for the control. The current state provides information concerning the assessment from our review, observation, or testing strategies. The outcome will provide detailed metrics based on a scoring or risk rating. The process will utilize a scoring category similar to the Inclusive Culture Toolkit.

Table 11 provides an example of a control gap report. The test results identify the program's current state, and the risk identifies the weakness or gap. Each program risk has a risk score assigned. For example, the O.1—Information Sharing Team Communication is rated "5" as the "Program Risk" score. Throughout the assessment lifecycle, the score will remain unless a new categorization is required. The investigator utilizes the information to analyze the overall gap analysis and "Potential Risk Reduction." The gap analysis process may take months in many organizations, so the example provided is designed to support familiarity.

Control ID	Control Name	Test Results	Program Risk
A.3	Communication Plan	The plan did not include notification timelines or incident response teams that require situational awareness data.	5- Without appropriate timelines, the reporting for incidents was late or inaccurate. 10- The IR team relied upon outdated data in making their decisions.
O.1	Information Sharing	Sharing between group A and the response team did not exist. The timeframe to share Level I incidents was not included within the information-sharing policy.	5- Team communication was obsolete and caused significant communication delays. 15-The sharing policy was not inclusive to the culture goals: share information so each unit can understand the current environment.
T.2	Notification System	An electronic notification system exists and is tested monthly.	0-No significant gap.

TABLE 11. CONTROL GAP

In Table 12, the scoring for each area has been identified and provided a "weighed factor." A weighed factor is a benchmark score that determines the control performance rating. In our example, the information sharing score was deficient. The control performance was weighed against the organization's "ideal" score and used to determine the overall gap. The weighed factors use an arbitrary numbering system, as outlined within Table 12. The Risk Score represents the scoring category, and the % of Potential Risk Reduction is a weighted factor. The investigator assigns the risk scores based on the test results (Performance). The risk reduction is computed through what was remediated. Do not confuse the score with the Minimum, Target, and Stretch.

Risk Score for SA Program	% of Potential Risk Reduction—Weighed		
	Minimum Threshold	Target	Stretch
Less than 5	75%	90%	100%
Between 5 and 9	75%	90%	100%
Between 10 and 15	70%	85%	100%
Between 16 and 20	70%	85%	100%
Between 21 and 25	60%	80%	100%
Greater than 26	60%	80%	100%

Control Functionality	Risk Score	Risk Reduction	Final Gap Analysis
A.3 Communication Plan	20	71%	29%
O.1 Sharing Information	15	84%	16%

TABLE 12

In Table 12, a "Control Functionality" has been created to describe the assessed control. In our example, the "Communication Plan—Risk Score" achieved 71 percent since not all issues were remediated. Basically, 71 percent of the problems are compliant; and 29 percent need remediation. If all issues were remediated, a score of "100" would be assigned. The "Final Gap Analysis" is derived from "100 minus Risk Reduction." The overall gap analysis for the program is calculated by doing the following:

Gap Analysis1 (29%) + Gap Analysis2 (16%) / Total Gap Areas (2) = Overall Gap Analysis (22.5%)

The overall gap analysis is used to determine how the organization can drive change in meeting its goals. These goals represent the ideal state or operational condition where the organization's mindset fully embraces situational awareness. A score of 22.5 percent can fall into many categories for approaching their gaps. One particular strategy is to create a rating based upon the gap score—such as Low, Medium, High, or Critical severities. Some organizations may describe severities as risk rating, security classification, or gap status. In our example, a final score of 22.5 percent would cause a low concern.

Low 0-25% **Medium** 26%—50% **High** 51%—75% **Critical** 76% -100%

The organization may review the gap analysis and create future remediation tasks. In the later chapters, there will be an advanced discussion on how a risk-based approach applies to this concept. The remediation tasks are designed to correct any discrepancies or deficiencies with the program. A more common reference such as the Corrective Action Plan (CAP), Plan of Action and Milestones (POA&M), or Vulnerability Remediation and Planning (VR&P) may be used to describe the task. At this stage, the organization is tasked to remove the security gaps or deficiencies. Let's continue the SAAF and focus on correcting the gaps.

For the next stage, the remediation tasks are the main priority. The remediation activity consists of corrective actions, due dates, and actions required to improve the program performance. Most organizations consider this a remediation plan, and as stated, a CAP, POA&M, or VR&P provides the same strategy. The overall objective is to correct security deficiencies or gaps. Organizations would generally assign the task to subject-matter experts (SMEs) for oversight. Typical work may include project management tasks or monthly remediation reports. In the tasks, the organization works to ensure the culture understands remediation goals, which aligns with the inclusive culture concepts. The monthly reports provide management with status, progress, and task accomplishments. Table 13 provides an example of a typical remediation plan.

Control ID	Program Risk	SME	Task Status	Due Date
A.3	5—Without appropriate time-lines, the reporting for incidents was late or inaccurate.	John Jones Lead IR Architect	08/20—3 of 5 IR guides updated	10/20/2020
A.3	10—The IR team relied upon outdated data in making their decisions.	Ken Roberts Lead IR Coordinator	08/20—Team training conducted	10/25/2020
O.1	5—Team communication was obsolete and caused significant communication delays.	Terry James IS Analyst	08/20—Held Table Top exercise	10/13/2020
O.1	15—The sharing policy was not inclusive to the culture goals: share information so each unit can understand the current environment.	Derrick Man IS Analyst	08/20—Updated IS Sharing Policy	10/22/2020

TABLE 13. REMEDIATION PLAN

The remediation plan stays active until management discovers that each program risk is remediated. The Final Gap Analysis score will decrease as the remediation activities occur. The end goal is to shorten the gap and increase the risk reduction. As the gap decreases, the organization's situational awareness profile will sharpen and model the culture image. Information sharing will occur and automate the communication plan and cause the organization to attack incidents at unprecedented speeds. Another outcome is that the partnering companies can gain trust that the information-sharing process functions—represents a buy-in structure. The information sharing and buy-in structure repeatedly and heavily depend on business relationships between many organizations. To establish a more substantial relationship, each organization may engage in continuous monitoring (CM). The purpose of CM is to mutually provide knowledge about ongoing security practice and situational awareness programs. CM aligns with ongoing assessment and provides early indications and performance status.

Continuous assessment (CA) provides reinforcement and systematic analysis of the environment. The phase identifies and realigns the situational awareness program. Many of the SAAF processes are periodically assessed and integrated into the continuous assessment cycle. The use of SIEM tools, performance factors, or regular reports is used to determine program effectiveness. In our example, trend analysis can support various performance factors and identify gaps or weaknesses.

The most valuable CA tools are data calls and action items. Data calls are periodic requests designed to solicit or collect data concerning the controls, such as the communication plan passing or failing scores. Action items are single tasks that require completion. The task could require a SOC team to update the communication plan periodically. Despite what is used, they both are valuable tools for the SAAF process. The outcome provides security control benchmarks, performance scores, or risk ratings. Organizations typically deploy both through a project management lifecycle and track their progress monthly. The result describes whether the operational programs conform to its standard.

An operational profile is a critical element toward having an effective situational awareness program and prioritizing remediation strategies. When specific operational tasks and policies are not addressed, they affect the ability to sustain security. When we address 360-degree visibility, we are essentially addressing an organization's security state and operational characteristics. Imagine not having complete information on situational awareness and its security gaps.

SAAF PROFILE QUESTIONS TO PONDER

- How many different security environments are on the enterprise?
- What situational awareness controls are implemented?
- How do external partners communicate through information sharing?
- What program threats and vulnerabilities exist?
- What situational awareness components does the organization address?

The most critical CA element is the reporting structure. The proper reporting techniques allow the assessment process to gather accurate and complete data. An organization can deploy data quality management tasks or quality checks to identify reporting gaps. The requirement becomes particularly important when involving information sharing and change management practices. External partnerships will lose confidence when "bad data" is reported. The change management process requires accurate information for automated tools since they provide real-time situational data and alerts. If SIEMs, vulnerability scanners, or asset databases are updated or changed, the notification must be shared with the correct person or business units. Typical reporting is described through electronic data or hardcopy information. However, other reporting lines such as email, status boards, or meetings require accurate data. As an evaluator, it's essential to have a deep understanding of the CA reporting process, and once again: "bad data in" equals "bad data out."

CONTINUOUS REPORTING BENEFITS

- Increases situational awareness and supports reporting requirements.
- Trend in overall residual risk, broken down by inherited risk, accepted risk, and risk to be mitigated by remediation plans and system-specific risk analysis.
- Top risk contributors by security controls and system components, status of open and remaining gaps or deficiencies.

In a typical environment, the CA process and its reporting techniques are automated or defaulted into the cybersecurity culture. Many organizations face challenges because the SAAF concepts are low priorities or removed as a security practice. All too common, the strategy forces additional risks. The blame shifting becomes a familiar path, and management teams must decide what risks are the most important. In the process, information sharing or communication controls become risk-prone. So what's the answer?

Management teams and decision-makers are faced with many responsibilities to build and deploy the most comprehensive situational awareness model and answer risk-based questions. The Department of Homeland Security has established CISA Central as a hub supporting accessing more data sources, designing tools to translate information and intelligence, and information sharing amongst the cybersecurity community. The end-state ensures that situational awareness components and the nation's infrastructure operate as a single entity. Will this resolve risks associated with situational awareness?

The Cybersecurity Mindset integrates the inclusive culture and situational awareness model. As previously discussed, these enablers promote cyber hygiene and a "buy-in" for achieving the Cybersecurity Mindset. The alignment happens since every technology culture is onboard and defines situational awareness as a business driver. The principal business units are actively involved in the process, and to further understand the entire model, they apply risk concepts. Involvement requires an approach where risk is understood, documented, remediated, and mitigated. To build a complete entity where the risk-based approach is successful, the inclusive culture, situational awareness, and risk mentality must align. Let's further the topic by discussing what risk-thinking and the Cybersecurity Mindset have in common!

• VIRTUALIZED PATH •

- Inclusive Culture
- Situational Awareness
- Risk-Based Thinking

RISK-BASED THINKING

In 2020, the COVID-19 pandemic shifted the technology landscape, business, and working requirements. The culture for remote working surfaced, and many organizations' preparedness was tested. IT environments scrambled to create remote service and teleworking initiatives for many employees, which caused service gaps, delays, and stress. The cybersecurity landscape was in dire straits, and some organizations survived while others became risk-prone. Also, the governance strategies were exposed—their risk programs were weak, and security gaps surfaced.

IT risk programs survive when the governance strategy onboards a winning culture. The core offerings must be inclusive—in that the entire security framework, practices, or management practices mirrors the corporate vision. The enterprise data, applications, infrastructure, and varying services must receive equal attention and safeguards. During the COVID-19 pandemic, the process failed. Many of the core security services were not positioned to support remote operations. The failures resulted from key security initiatives and attitudes being culturally separated. The security functions operated with a preventive mindset—let's secure the environment before remote onboarding operations—and the entity focus was the customer's first. Trying to convey security was very foreign to business operations. Within every risk culture, management and functional teams must balance security and business. As with COVID-19, there existed an imbalance for risk planning and thinking cybersecurity, which generated security risks and service delays.

The responsibility for COVID-19 technology and cybersecurity governance failures was evident. One particular area was emergency operations and disaster preparedness. In every IT environment, a disaster recovery plan should exist that describes its emergency operations. It contains an employee work location, remote support conditions, and system availability standards. When a disaster strikes, these security functions support emergency operations. If the processes fail, the outcome poses significant risks, and this defeats the purpose of having a risk culture.

An entire culture can live and breathe risk management when risk-based thinking exists. It provides an outlet where risks are assessed and managed through a framework or lifecycle. The action is automated, and the security culture speaks "risks" and uses protective chains to secure the environment. It's defined as a secure link that safeguards information assets. Examples are authentication, isolation, and authorization. Each particular link is technology-connected and serves the end-state: *let's protect the enterprise.* For instance, access control and remote operations supported teleworking service during COVID-19. When developing the teleworking strategy, organizations had to integrate authentication standards and virtual private networks (VPNs), the primary security technologies required to execute a remote log-on session.

Risk-based thinking is an organizational-wide effort to examine and control risks for every platform, service, operation, or connection. Corporations must wear a "risk hat" and establish a buy-in culture where situational awareness, growth mindset, and other cybersecurity principles are connected. As previously stated, the Cybersecurity Mindset is a connected methodology. The culture, situational awareness, and risk mentality co-exist as business drivers. They are all encapsulated into the mindset model to advance security practices—such as a holistic defense practice.

The term "holistic" describes risk thinking as a complete system, and it helps organizations manage their entire governance strategy. All the intricate practices, decisions, engagements, and operational requirements are designed specifically for the culture. In short, risk, inclusive culture, and situational awareness are enclosed programs.

In **Table 14**, management, operational, technical, and service-related functions layout the security architect. Depending on an organization's security architect, there could exist similar or more risk-related areas. Despite which technology platform exists for the IT architect, the risk mentality is required and relatable across each functional area. It's a whole culture that considers risk from every angle, crevice, business objective, or strategy.

Management	Operational	Technical	Service
Budgeting	Audits	Database Storage	Applications
Cybersecurity Programs	End-Point Protection	Encryption Technologies	Business Process
Governance	Incident Reporting	Network Configuration	Data Visualization
Strategies	Threat Detection	Security Engineering	Scalability
Training	Vulnerability Management	Technology Integration	SLA

TABLE 14

The risk mentality is measured according to performance factors and the security state. Every corporation's technology profile includes its current vulnerabilities and threats, which derives the risk picture (Threats x Vulnerabilities = Risks). Having the mentality and approach to understanding the formula carries weight. Within the security mentality, a person thinks about the big picture—how safeguards, protection standards, and organizations survive the hacker's appetite. One survival technique is to transition compliance-based frameworks into a risk process. The concept examines compliance vs. risk initiatives and isolates any security failures. Organizations align their defensive mindset with combating security "outside the box."

CHAPTER ELEVEN

EXTENDING A COMPLIANCE MENTALITY

Through various client engagements and learning the cybersecurity landscape, I have experienced multiple approaches to risk management. Each project used independent concepts to evaluate its programs. Some of the techniques were heavily constructed around budgets, while others were purely risk-driven. The work-related functions were closely matched, but the outcomes were separated into compliance or risk. By default, the security teams followed the policies and procedures to address their tasks. Their involvement consisted of various validation checks and assessments that supported benchmarks. Although the functions were satisfactory, the outcome created additional burdens and security failures. One is the discovery of vulnerable actions. In the deep crevices of the environments, there existed significant loopholes. These loopholes exposed weaknesses and deficiencies for configuration settings and vulnerable applications. The outcome was purely descriptive—Threats x Vulnerabilities = Risks—and compliance provides overall indications for readiness and future remediation tasks.

Risk management and compliance are connected. In many organizations, they are treated differently, and the outcome induces additional complications and security gaps. When shaping the enterprise strategy, it's more efficient and successful when both are integrated into the risk mentality model. For instance, various laws and regulations may mandate operational measures, and employees must follow the laws. If they are followed, the organization can indicate its status as compliant and think risk. It's beyond a "check-in-the-box" to strategize a risk management program. A deeper analysis of whether the laws operate is more sufficient. It provides a detailed image of the operational characteristics, risk-thinking ability, and security gaps. Despite the strategy, procedure, or validation used, compliance and risk operate separately and support each outcome.

DIAGRAM-3: GRC LIFECYCLE

The failure for risk initiatives stems from governance teams practicing more compliance than risk. These groups are more strategic than tactical at describing security requirements. Governance teams' goal is to drive a strategic plan and identify long-range visions. These visions are to ensure certain performance marks, validation checks, or goals are met. Although the process sounds beneficial, there are significant drawbacks. The entire security architect is benchmarked to achieve a satisfactory score. These scores are self-developed and serve as milestones. For instance, the performance score of 90 percent represents a 10 percent gap. The governance framework can establish benchmarks where a consistent score of 90 percent within 90 days is satisfactory. Wait! The word satisfactory represents the minimum-security requirements. So, what happens to the 10 percent? Since the governance team is happy about 90 percent, it's overlooked. The entire culture is at risk, which results from compliance failures. Sometimes "box-checking" can cause more harm!

The box-checking is a compliance initiative that drives certain milestones or performance points. Its use scores and standards as security indicators and corporate security objectives will use the information as success factors. Additional objectives may describe the checks as the end-state or security goal. Despite which strategy is used, the risk hat must be worn. Every environment has its drawbacks and standards—so box-checking can serve as a preliminary indicator. It allows the culture to design frameworks, interpret, and treat risks. So when examining the inclusive culture, the growth mindset can transition box-checking into a risk-based culture.

One of the most widely used box-checking objectives is patch management. The NIST Special Publication 800-40 Revision 3 Patch Management—Guide to Enterprise Patch Management Technologies states, "Patch Management is the process for identifying, acquiring, installing, and verifying patches for products and systems. Patches correct security and functionality problems in software and firmware." It's a classic and relatable definition for patch management. The program utilizes monthly scanning reports to indicate metrics, host count, aging summaries, or severities. These numbers also provide actionable data on the program strategy. Typically, organizations use tools such as Microsoft Endpoint Configuration Manager or BMC Remedy to deploy the patches.

So how does the box-checking really work? Let's say a vulnerability report identifies 25K updates with a mix of servers, desktops, third-party applications, or security updates. To drive compliance, the GRC team includes that a desktop vulnerability remediation score of 90 percent is compliant. Their reasoning could stem from various mandates and deadlines—so achieving a 90 percent as a "box score" would suffice. Next, the patch coordinator plans and executes the patch deployment cycle; afterward, they submit a compliance score. In this example, let's assume that 5.6K of the 6K desktops were fully patched. The 400K for the desktops and the remaining 19K of vulnerabilities need attention, and here is where the disconnect occurs. The complete focus should be 25K of vulnerabilities and tactics. Patch teams' focus is driven toward governance strategies and not risk-related thinking—so the 19.4K of vulnerabilities are at risk. Let's think tactical!

The tactical approach drives a different concept, as it's a Tier II initiative. The thought process is more analytical. As would risk management, you cannot omit compliance and risk, and injecting a tactical approach considers the entire picture. Every security indicator, gap, or severity is analyzed against probabilities and included within the remediation plan. Typical terms such as a risk analysis will deep dive into the controls and highlight actual risks. The cycle consists of analyzing threats, vulnerabilities, and controls to determine fundamental weaknesses. As with the strategic approach, the process operates from the surface: *how many checks are completed?* When referencing a benchmark score of 90 percent, the tactical approach examines the 10 percent gap to determine the severity or impact.

A more concrete relationship for tactical engagements is holistic defense. As stated, before holistic defense considers the entire architect. In a tactical engagement, the security controls are extensively analyzed, and all vulnerabilities are remediated. The approach provides more insight and transparency into the risk-analysis process and defending enterprise assets and information. The Department of Defense (DoD) uses a similar technique called Defense-In-Depth, a layered defense model. Each layer provides a protection standard, and as security travels toward the core, the controls become tighter. The concept includes varying strategies to secure traffic and entrance points from internal or external sources. The model also provides adaptation principles where agencies can address on-demand changes and configuration requirements. When examining risk protection and its mentality, the model is very holistic—as it encompasses the whole security landscape. Every security function is viewed before the system becomes operational, and each security function is connected and addresses specific needs. The DoD Information Assurance Technology Framework (IATF) provides the following in defining some of the Defense-In-Depth Model concepts:

- Adhering to principles, commonality, standardization, procedures, and
- interoperability.
- Judiciously using emerging technologies and balancing enhanced capability with increased risk.
- Employing multiple means of threat mitigation, overlapping protection approaches to counter anticipated events.

There are many tools and customized reports in today's IT environment to visualize security trends and metrics. The indicators provide status, changing events, or performance indicators. The involved personnel or teams can communicate false analysis, which implies that the diagnosis was wrongly interpreted. A more common term expressed is false positive or negative. When a false positive occurs, the indicator displays a failing status when it's a success. With the false negative, it's the opposite. An indicator will display a success status, but the underline condition is a failure.

When it comes to compliance and visualization tools, it's essential to consider integrating situational awareness, false analysis, and compliance. These indicators require environmental knowledge and understanding of the holistic defense model. Compliance is not a failing contributor, nuisance to technology, or the all-inclusive answer, but we must extend its process into risk-based concepts. For instance, a web-visualization tool that provides graphic indications can represent compliance by having trending bars. The viewer may interpret the results as a positive status. In our example, let's assume the system encountered a software deficiency that provided incorrect data. The viewer would still consider the indications as a success—and the underline condition is a false negative. When the user reports the compliance status, the environment may feel secure. If the holistic defense model were practiced, the technology teams would note that the visualization tool provides false indicators. Through using the model, they would investigate and question the results. Again, situational awareness and inclusive culture principles are required. Even though a compliance strategy, they navigate organizations to think risk and share information. It's common for teams to share inaccurate compliance data, so again, our defense-in-depth or holistic defense model becomes essential.

The holistic defense model was experienced as an information assurance (IA) analyst while working for the Department of Defense. The projects practiced and enforced Defense-In-Depth through security integration and assessments. As an IA analyst, the experience created a comprehensive view into compliance and risk and security diversification. In some similarities, security diversification models the holistic and defense-in-depth model—as it encompasses the entire scope, process, picture, or risk engagement. In its nature, security diversification is an investment.

The methodology extends compliance, balances risks, and promotes risk thinking. The entire security scope is validated, tested, and analyzed beyond the compliance mentality. It's similar to mutual fund investments. Every investor develops their portfolio to balance risk, which means they choose funds from various sectors such as US Index, Mid-Cap, Small-Cap, or Bund funds. Each fund operates in different markets. As with security diversification, it's about selecting, integrating, or using various security practices for the security architect, so the portfolio and architect are buddies! Each security practice extends compliance to a risk-based environment. When a team views its incidents, vulnerabilities, asset counts, or remediation closure, they think beyond the numbers and visualize the architect as risk-based. The same exists with the mutual fund portfolio—investments think beyond the numbers and integrate risk-based decisions. How is your security investment today!

Visual indications or graphical analysis has become a handy technology tool. As risks become more apparent, the technology culture needs information and just-in-time data. The underneath operations or data-mining techniques are essential, but they can become problematic from a risk-mentality perspective. One area that is far more problematic is "color indicators." These are practices that enjoy seeing "green over red." When reports are distributed, the commonality shows success rates, benchmarks, or milestone percentages in colors—green passes or red fails. Another term is visual compliance, which also contributes to decisions and indicators.

In 2012, I started a project providing vulnerability and remediation support. The project required common vulnerability taskings such as scanning, reporting, and metrics development. Daily our team would analyze reports and extract key metrics for management. We had first-hand knowledge of risk conditions and color indicators. There were many instances where compliance and risk initiatives were separated. One was that the government created in-house rules that every report must work toward "The Green." Let's step back for a moment! The green was a compliance initiative that separated risks—so why would management impose a policy? The gap in understanding or onboarding the risk mentality is the answer. It's not proclaiming the government was pursuing the wrong security path. However, it does reference a key point, the risk-based approach was undervalued and not prioritized. Daily, we would communicate side jokes about receiving "traffic tickets." The idea represented that our leaders would impose working restrictions or punish the team. For instance, a report may indicate an upward trend (10 percent current month) and a downward trend (25 percent the previous month) for Windows-based vulnerabilities. The indications would yield a "red" for the current month, and then management would start issuing "traffic tickets." I remember the meetings, emails, and justification delivered to defend the outcome. In many of the reports, we would describe our risk-based process and root-cause analysis. Some of the information presented would state that new patches, new computers, or further signature checks were the solution. Despite the efforts, the management team still implied, "We must see the green," which means the risk connection was less concerned. Many of the team leaders attempted to sell the "risk" so management would connect the security picture. As usual, the outcome presented a reoccurring process: "green is good." The point to consider is that compliance initiatives can misguide security when benchmarks or visual indicators are prioritized. It's beyond the color spectrum or high-end graphics—we must examine the root cause and deliver solutions that merge compliance and risk.

Another area that focuses on compliance and risk is key risk indicators (KRI). It serves as an enabler in determining risk areas that need attention and remediation. Organizations can draw valuable data that describes the predictability or possible risk signals. Early indications provide direct links to the enterprise risk appetite, which is the risk-acceptance level. Compared to compliance, the indications drill down into the root cause and the holistic process. Another advantage is that the timely decision-making process is accelerated since the research is done early. In the cybersecurity community, KRIs are used quite frequently to predict security deficiencies and actionable intelligence. When referencing the "green," the KRI can advance the visualization into actionable and root-cause decisions. The "green" can become a barrier since management may absorb the concept of a safe zone. It's somewhat similar to a traffic light—when we see green, it's safe!

The brother concept for KRI is the KPI—key performance indicators. The KPIs are benchmarked metrics that describe the performance state or quantifiable measurements for the security performance. The environment may integrate factors that track the number of vulnerabilities remediated over thirty days. The results are compared against a standard score such as 85 percent, which means that 85 percent of the total count must be remediated. Again, the 15 percent gap has surfaced! Should the team stop at 85 percent and cross their legs or invoke root-cause analysis and gap closure? Imagine a visualization platform that displayed the word complaint in prominent bold characters. The text would direct the team to say, "We are in the green," and possibly celebrate over a long lunch break. After lunch, the problem started because 15 percent represented a major malware flaw. The critical point is that KPIs represent numerical values and benchmark goals. If the KRIs are integrated with the KPIs, the security action is balanced, efficient, and provides better protection. Each KPI is further evaluated through the KRI procedure. The 15 percent factor is automated procedures. For example, the security gaps are assessed through a holistic approach, and the malware flaw is categorized as a high-severity. KRI uses previous risk analysis and root-cause data, which references the vulnerabilities as a high-impact. The actionable intelligence, industry trends, or impact assessments correlate to the KRI high-impact vulnerabilities and performance status. This is where the KRI and KPI process connects and provides true-risk indications.

Beyond the KRI and KPI scope, there exist processes that contribute to the risk mentality. The governance and compliance strategies were briefly discussed and brought to the surface some valuable information; compliance and risk are mutually dependent. An extended discussion brings to attention security frameworks such as the National Institute of Standards and Technology (NIST)—Risk Management Framework and internal Governance, Risk, ISO IEC 27001/ISO 27002, FedRAMP Governance, Risk, and Compliance (GRC) programs. Each framework has its pros and cons and offers concepts and processes to extend the compliance mentality. Every agency or organization onboards the frameworks engages compliance, and the frameworks extend the benchmarks into a risk-based environment. For instance, the NIST framework operates through the SDLC cycle and uses validation testing to determine weaknesses or deficiencies. The outcome provides relevant facts and performance factors about the system or application security accreditation status. The federal agencies extend the compliance initiatives and work toward remediation. When each application or system receives its full accreditation, the environment moves into a continuous monitoring phase. During this phase, compliance and risk are well balanced. The agency performs validation checks and remediates security deficiencies through ongoing evaluation and testing. The security controls are constantly evaluated through an application, code, or vulnerability scanning. The defacto tools such as Tenable Nessus, Fortify, or DISA STIG benchmarks are used to validate the compliance and risk posture. Various reports are distributed and updated throughout the system lifecycle, and included are compliance and risk-related status.

There are various GRC tools on the market to assist with security assessments or validation checks. Each platform from RiskVision to Archer provides the relevant compliance picture and risk indicators, and always, the use of the tools is where risks are either lapsed or remediated. It's not an indicator that the devices are dysfunctional. However, it serves as a true statement that GRC tools speak value when used to the full extent. The technology industry is a number-based culture, which is great, but extending the compliance mentality and risk integration is more beneficial. Each GRC tool onboards compliance as a strategic initiative and deploys initiatives such as operations and tactics. With operations, the business units carry out risk missions and goals. Their operational risks are evaluated to support the customer or end-users. The tactical approach addresses the risk mitigation and balances outcomes to strategic objectives—let's ensure we are compliant. The KPIs and KRIs are onboarded and driven to reduce enterprise risks through different tactics or remediation procedures. The GRC initiatives are a well-balanced and efficient tool for achieving compliance and extending its outcome into a risk-mentality. The diagram below provides a high-level GRC concept and methodology. Each functional area addresses governance, risk, and compliance as a framework.

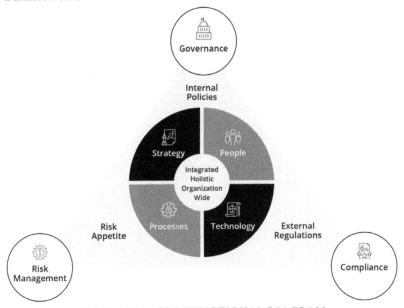

DIAGRAM-4: GRC FUNCTIONAL DIAGRAM

The process of extending the compliance mentality brings forth a whole security picture and its functional requirements. The separation in understanding compliance and its benchmark has been one failing factor. However, the root cause extends into the risk mentality. Most IT organizations practice security from the surface, which means the Defense-In-Depth principle failed or was categorized as a low prioritization. In the discussion for "Extending a Compliance Mentality," many pointers and methodologies were communicated to promote compliance and risk as a single initiative. How corporations and business units use the concept provides better results. The results demonstrate whether the environment mitigates risk and drives a successful risk-thinking program. In the security culture, these factors connect to assessments and help cease opportunities. In this place, teams discover dangers and develop improvements. Sounds like the growth mindset! Let's see how risk discovery leads to success.

CHAPTER TWELVE

RISK DISCOVERY AND OPPORTUNITIES

The IT workforce has experienced its challenges and transitions. Since the pre-historic days of computers, the industry developed motives to strengthen technology and its integration—part of the process involved learning efficient methods to become innovative. The concept further elaborates into IT professionals' working habits and objectives. They contribute an enormous amount of time, energy, and resources in resolving complicated and unconventional issues. Within the entire process, there exists moments or periods where they reinvent or create solutions outside the box, which means a book solution or technical guide cannot answer a specific requirement.

During the naval years, our technical approach modeled the outside-the-box methodology. There were many situations where regular maintenance was scheduled outside the routine schedule. In the Indian Ocean and the Mediterranean Sea, the systems would require the most critical care due to ambient temperatures or power fluctuations. As technicians, we took advantage of the time to perform major repairs and change defective parts—considering that it was impossible during system uptime. Many circuit cards, power supplies, and conversion units were replaced between system reboots or power outages. For instance, a communication halt would warrant a system reboot; and the technicians would utilize the period for replacing parts and defective modules. The period became a success due to redundancy, which created backup options. As technicians, we believed that the backup option saved lives and "ceased the opportunity" in keeping the ship moving.

The connection between the US Navy experience and risk-based thinking carries similarities. They both align with seizing the opportunities and taking advantage of options. As with both, no organization can prepare for every situation and plan risk-based opportunities. In the cybersecurity environment, technical teams and managers are no strangers to the concept. Every security assessment or audit may produce weaknesses or gaps. Within each, there rely upon advanced chances to understand risk profiles. While some cases are successful, a community of overlooked issues where risk discovery and opportunities (RD&O) become essential will exist.

Once risks are discovered, the initial fear of job loss, embarrassment, or additional work can surface. Every IT organization protects its image through silence and limiting information sharing. The strategy is extended when risk impact and misunderstandings occur. Risk discovery provides the opportunity to understand security performance and overall maturity. These are points that describe cyber exposure and define what risks exist. For instance, the disclosure of open malware ports may identify protection status and configuration gaps. Businesses can use the data in developing preventive, detective, or corrective measures. Some corporations think their organizations are hacker-proof and risk resilient—so they may overlook security performance or open malware ports. For instance, when we usually hear that a company has never been hacked, the next question assumes whether they are actively monitoring their enterprise. Usually, hacks are discovered when monitoring a system—so no monitoring equals no indications. Also, such a company may not onboard periodic checks and assessments, identifying areas and programs requiring cyber care.

The term "cyber care" addresses risk mitigation, maturity, and remediation. When cyber care is applied, organizations develop avenues and roadmaps to safeguard their enterprises. The security failures, threat factors, and gaps serve as opportunities. How businesses utilize the options determines their ongoing risk posture and capability to safeguard information assets. The common saying "you never know until you know" becomes apparent. Cyber professionals are *only as good as they inspect*. If a subpar inspection occurs, there will exist sub-par results.

The risk-fear factor is standard when environments are exposed. It helps businesses understand that cyber hygiene can prolong a healthy security program. It's somewhat similar to health hygiene—our health condition is only known when we conduct doctor visits or gain a specific illness. Our goal is preventive care, so practicing proper hygiene and regular doctor visits can prevent diseases and provide early health warnings and conditions. With cyber hygiene, the assessments and checks are preventive strategies businesses use to discover or reduce enterprise risks. The indications and outcomes provide early warnings and advanced information on defending enterprise assets and data.

One professional group that practices cyber hygiene regularly is the risk and vulnerability management (R&VM) teams. The R&VM group exhausts their time approaching risk at the front line—a simple metaphor that describes the first level of decision-makers. These are the gatekeepers or the first visible group to engage in security risks. Their decisions determine the security path and enterprise engagements. As a junior security analyst, I experienced many situations where the "gate-keepers" did not seize opportunities. There was more shifting-the-blame vs. thinking risk. The practice evolved into more risk-related issues than solutions, and the root cause was more cultural-based than management decisions, and it's relatable to the inclusive culture. The outcome is resolvable by removing barriers that exclude the risk-mentality and associating a culture where risk, opportunities, and success operate as a single entity. Another alignment is situational awareness. The entire environmental knowledge would focus on risk-based approaches. That means each involved personnel would think and utilize risk-based decisions. As stated in the previous chapters, the Cybersecurity Mindset is a holistic approach and process—you cannot have an inclusive culture, situational awareness, and risk-based approach operate as separate actions.

When leveraging my career path, there were many situations where opportunities were overlooked or excluded from risk management procedures, such as audits and assessments. Many of the evaluations provided early indications and warnings for a potential issue. Teams involved would discover innovative and clever avenues to sell a familiar pitch: the department "jacked up" is responsible for the finding. The more the audits and assessments occurred, the more they enacted upon their clever ideas; and thus, a revolving door prevented the opportunity for success. Still today, there are probably some teams and corporations that practice the same strategy. Does this mean the risk-based approach is a dead issue? No, because it's more overlooked than processed into technology taskings and requirements.

One of the most common opportunities for success is the Federal Information Security Management Act (FISMA) audit. The federal government mandates that agencies conduct their audits annually by a perspective Office of Inspection General (OIG). It was enacted as part of the Electronic Government Act of 2002, designed to enhance the government's electronic service and information technology protection. FISMA gains success by protecting information from unauthorized access, use, disclosure, disruption, modification, or destruction. In 2014, FISMA became the Federal Information Security Modernization Act of 2014—and still served as official law.

FISMA is successful at providing the most current risk-based opportunities and cost-effective security. Each audit provides high-level recommendations and maturity milestones, and the audited agency is required to enact and remediate the findings. The security controls follow NIST-800-53, a comprehensive list of federal security controls. The process is straightforward and leverages a risk-based approach. The Office of Management and Budget (OMB) uses the combined results for congressional reporting. Congress can use the information to determine the country's security posture and the improvements required. Although it's a mandated law, the FISMA audit automates the risk-based opportunities. When audits are not ordered, there exists a significant gap in approaching opportunities. The engagement is organizational-dependent and cultural-based, and freedom is exercised on what "matters most." The statement does not concur that automated means are advantaged, but it highlights that automated risk-based approaches are well-structured and designed, and the opportunities for success exist. Additional audit programs such as the SOC (System and Organization Controls) audits—SOC 1, SOC 2, and SOC 3—carry similar processes and highlight relevant security gaps. Corporations use their outcomes to drive a risk-mentality and foster improved security measures.

Assessments provide similar outcomes as audits. The difference is that they are not law-driven, so the approach follows specific frameworks. The review is an internal standard, while the audit follows external standards. The term external means the business or agency follows a third party or governing guidelines, and they are structured to validate many security settings and operational processes. A typical evaluation will examine various web-based infrastructure, applications, or mobile devices for security weaknesses or vulnerabilities. Although the assessments are not as comprehensive as an audit, they still provide risk-based opportunities. Validation occurs through using automated scanning tools such as Fortify, Qradar, Qualys, or Tenable Nessus. These tools examine specific file locations and root folders for software settings, port status, malware indications, or misconfigurations. The advantage is that the results identify vulnerabilities before audits or serious security violations occur. Table 15 outlines key areas that are usually assessed.

Technique	Capabilities
Network Discovery	• Discovers active devices • Identifies communication paths • Facilitates the determination of network architectures
Network Port and Service	• Discovers active devices • Discovers open ports and associated services/applications
Vulnerability Scanning	• Identifies hosts and open ports • Identifies known vulnerabilities • Often provides advice on mitigating discovered vulnerabilities

TABLE 15. ASSESSMENT METHODOLOGIES

Risk-based opportunities operate well within an assessment strategy. The evaluations provide actionable insight into discovered vulnerabilities and their impact. A business or federal agency may use it for information advantage and early warnings. When executing the assessment, there are ample opportunities to discover overlooked security gaps or broken safeguards. Since there is either high tempo or advanced priorities, the RD&O can become less critical. Alternatively, when options are labeled a priority, organizations can promote cyber hygiene and discover the unknowns. It's a cyber hygiene practice since it provides the preventative process to engage risks and remediate their impact. The diagnosis and outcomes are indicators, recommendations, and remediation opportunities. A successful network scan will identify risk exposure, and the affected system can be remediated or mitigated as a preventive measure. Here are some questions to consider when designing a risk-based mentality:

• How might the vulnerability open the door for hacking?
• Will the vulnerability impact the risk posture?

- What systems are affected and their criticality?
- Will this affect business operations?

When reviewing the questions, the entire strategy transitions into the holistic defense mindset—where are we affected across the enterprise, and how can we prevent risks from escalating? Let's assume an organization did not consider the RD&O and questions. The entire organization would work within the advisory stage and perform corrective actions. Sometimes the corrective actions add additional labor, time, and resources. It forces teams to operate in an ad-hoc mode and perform on-demand security taskings. In turn, many security gaps will occur, and now, the entire environment becomes a hacker's appetite. The preventive measures are more structured and aligned to identify and secure the environment before advisories. As a cyber caring individual or team, the practice can shorten the labor, time, resources, and on-demand requirements. In the Navy, the term ad-hoc meant "shooting from the hip" or being unstructured at performing tasks. The risk-based opportunities occur at all stages—so structure and alignment channel the process far better than ad-hoc operations.

IBM published its annual Data Breach Report in 2020, which was a comprehensive review of international data breaches and their costs. The report summarized various implications and risk-based shortcomings that existed when businesses' data protection standards were weak. According to the report, organizations that lacked security automation and an effective incident response program had higher spending costs; and 80 percent of the organizations experienced PII breaches. When referencing the cyber hygiene practice, it's apparent the affected companies did not consider preventive security measures, cyber care, or proper cyber hygiene.

A data breach exists when personal information—name, address, email address, social security number, driver's license ID, or financial information—is stolen or exposed. They occur due to some apparent lapse of security misconfiguration. The gap surfaces from weak passwords, unauthorized access, software vulnerabilities, or misconfigured infrastructure devices. Another factor is response times. If a violation occurred and was undetected within twenty-four days vs. twenty-four hours, the difference could cause security implications. One risk-based approach is to deploy security orchestration and automation technologies that ultimately improve detection and response times. These technical advances develop opportunities, drive early warnings, and promote predictive analytics. The manual labor, ad-hoc responses, and lapse in remediation timelines are shortened. Corporations gain early advantages when risk treatment strategies are required, preventing security breakdown between upstream and downstream responsibilities.

Data breaches are a security nightmare. Beyond its implications, there exists an abundance of security flaws and industry security concerns. In its crevices, ransomware, vulnerable applications, or phishing can disrupt business operations and affect costs. Early detection and preventive strategies can reduce the risk implications and costs. One measure that addresses early detection and combines preventive and detective security measures is penetration testing or pentest. In a traditional environment, the vulnerability teams execute network, applications, or web-based scanning to identify security flaws. Since the operation is internal, the risk-discovery phases may partially view the security openings. When pentest are deployed, third-party firms perform a very detailed validation against web services, infrastructure devices, databases, user access, social engineering, or end-point security. The entire pentest is driven through business requirements, such as needing to understand how vulnerable users are vulnerable to phishing campaigns. Unlike a vulnerability assessment that tests for vulnerabilities, the pentest exploits vulnerabilities. The opportunity provides a complete view of the risk posture.

Although the pentests provide early detection and an RD&O, their costs are challenging. A very detailed and high-quality pentest can cost between $15,000-$30,000, depending on the size, scope, complexity, experience, and methodology deployed. Smaller companies may not have the budget to conduct a pentest, so they rely on their vulnerability management programs. Larger firms have more flexibility and operational funds to cover a comprehensive pentest. Some firms charge by IP address, which can drive costs. For instance, an organization with 100,000 IP addresses could be charged $1.00 per address—the total cost equals $100,000. Imagine five pentests conducted annually! The opportunity for success increases and carries more weight than the costs, as a security incident may cost millions. The return on investment (ROI) is the end goal. This is where successful opportunities and risk-mentalities collaborate. They provide an avenue to outweigh costs and focus on what matters most: protecting resources, data, information, and applications.

One of the cross-relationships shared with penetration testing is digital modernization. The overall relationship establishes growth strategies where the risk-mentality promotes business development. Before RD&O exist, most environments are risk-prone. When opportunities surface, the security environment can reduce risks and mature. The initiative maneuvers risk into a new climate where digital modernization transforms organizational growth and operations perform without interruptions and sustain availability.

Digital modernization defines the process to adopt, transform, modernize, or develop new and enhanced cybersecurity measures. It allows businesses to incorporate risk-based approaches and defend their technologies. When assessments, audits, or penetration testing are conducted, its outcome provides maturity status. The indications direct businesses on where risk treatment should exist and weaknesses within its security fabric. The treatments offer indications for infrastructure, networking, web applications, and operational status and determine whether data protection standards require improvements. This is where growth opportunities formalize, and risk-conscious organizations will advance cybersecurity.

The government has been a significant contributor and driver for digital modernization. Every internal organization must follow congressional laws, standards, and practices and operate with a risk-conscious mindset. Seeing, thinking, and living securely is their fabric, and it survives because successful opportunities and digital modernization operate in parallel. Each agency achieves seamless and secure data interoperability across federal and commercial partnerships and integrates the most stringent data protection standards. The methodology addresses risk and security growth as their primary enablers. One agency that exemplifies the approach is the Center for Disease Control (CDC), as it has 3,000 industry partners and sustains a data-centric architect. The CDC's mission is to identify and define preventable health problems and maintain active surveillance of diseases through data collection, analysis, and distribution. Underline operations utilize many platforms and technologies for data protection and information sharing. The RD&O captured allowed the CDC to produce scientific and actional data concerning public health issues and standards.

In 2014, the CDC established its Public Health Surveillance and Data program to monitor, control, build, and deploy data amongst different units, partners, and federal agencies. The program models and develop resilient controls and advanced safeguards for their enterprise assets and information. Within the CDC's technology core, data security advances early detection into risk discovery and vulnerabilities. The agency operates its public data across many partnering agencies and platforms, such as internal audits, data-sharing policies, and assessments. As the agency matures and modernizes, the entire technology landscape becomes secure, trustworthy, and reliable. It end-state drives public health surveillance, research, and, ultimately, decision-making; identifies weaknesses, gaps, or failures; and transitions the results into opportunities. The non-security mindset may interpret weakness and gaps as negative situations. However, the security-focused attitude defines the situation as real-time awareness—and a roadmap toward digital modernization and cyber hygiene. According to the CDC Public Health Director, "The CDC must work more broadly as an agency to confront challenges and embrace opportunities that arise as partnerships, processes, data, and technology progress." The statement clearly articulates that RD&O is at the agency's forefront.

One of the most critical outcomes for risk discovery is transparency. To sustain the practice requires data management and 360-degrees of security visibility. Within many similar or dissimilar businesses to the CDC, gaining clarity allows current and future growth. A business can remediate and reduce weaknesses and prepare future requirements. In the context of strategic planning, the future state is a planning phase that starts early. The familiarity aligns with predictive analytics and preventive care, in that its goal is to strategize and operate cyber hygiene. The early detection and warnings allow corporations to gain transparency and early indications concerning security impact and what technologies require adjustments. As one can see, the function for transparency is clear: provide complete visibility.

Transparency is somewhat similar to a vertical wall. When the wall is standing, businesses cannot gain insight into their future state. Alternatively, when the wall collapses, it bridges transparency, and now, the company can focus on its fabric, define what matters most, and what safeguards, changes, processes, solutions, or cyber hygiene practices are required. When these enablers are successful, corporations can become proactive in security defense; this is where the frontlines for security thinking formulates.

The security frontlines are an aggressive and forward-thinking strategy that enables early prevention and detection: the profile models many wartime tactics and maneuverability strategies. The battle commander utilizes forward-thinking to evaluate risks and opportunities. As the campaign escalates, they become aware of their surroundings and what tactical decisions make sense. In the technology space, the same concept applies. Technical teams formulate their strategy from the front and include every risk factor and its discovery in their decision-making. As issues escalate, they can deliver informed decisions and discover risks and opportunities. A much closer term is proactive security measures and alternately reactive measures operate in an emergency or ad-hoc mode.

Operating through reactive measures promotes overlooked opportunities. For example, a vulnerability that has aged sixty days is far worse when it's ignored. The indications, reports, and gaps may present inaccurate or risk-prone data. The proactive state produces early signs and active security measures while the reactive state performs "catch up." It sounds like an excellent discussion for proactive and reactive security measures!

CHAPTER THIRTEEN

PROACTIVE AND REACTIVE MEASURES

The term forward-thinking applies to cybersecurity planning and developing strategies. Technical groups use the methodology in developing project plans, future requirements, or anticipated changes. The time expended planning, forecasting probabilities, costs, schedules, and tasks can become overwhelming. Managers contribute countless hours attempting to predict future requirements, and for various reasons, the plan fails. To address the issue, management may perform ad-hoc remediation or emergency corrections to sustain operations. Although the process works, it cannot serve as an all-inclusive strategy for cybersecurity planning. There are induced and inherited risks that could occur, so operating in the ad-hoc or emergency mode is dangerous. Eventually, the cycle restarts—then the fixation surfaces. Sounds all too familiar!

In my early IT career, our team worked long hours to sustain printers and desktop operations. The success relied upon forward-thinking and planning. There was little room for errors, as it slowed progress and caused significant delays. The entire command supported over 1,500 users and 800 printers, which our team configured, deployed, repaired, and inventoried. One planning strategy that kept the group ahead was to anticipate additional taskings. During service calls, we carried additional tools and printers to address unplanned or unordinary changes. Many situations existed where ad-hoc trouble tickets or critical maintenance requests surfaced. In our typical fashion, the team responded and corrected the tickets.

The cybersecurity environment has brought upon similar experiences. As a vulnerability analyst, I experienced various remediation taskings that surfaced from advisories or management demands. It seemed that after every vulnerability scan, the entire enterprise played "catch up," which was driven through on-demand requirements. For instance, the monthly reports were executed at the same intervals and required similar metrics and reporting standards. Upon delivering the information, there always existed one manager that complained about his environment and risks. To reduce the complaints, we conducted mid-month scans and communicated the remediation requirements early. The strategy reduced many vulnerabilities and kept leadership happy. Although the changes were successful, they created additional labor and challenges. Even today, the memories and labor-intensive work serve as a learning experience. One was that forward-thinking positions individuals or projects to operate ahead. No environment wants to operate backward and induce project failures. The entire outcome can affect positions, corporate goals, or technology risks. It's an everyday challenge to combat technology risks, so adding management risks expands the workload.

A perfect example is the vulnerability tasks that occur monthly. If management created more demands, that means the normal taskings would become risk-prone. The time, labor, and effort would stretch and induce weaknesses or gaps.

Over the years, there have been many solutions to counter on-demand challenges and ad-hoc scheduling for security taskings. Some groups have focused on adding more tools or resources or changing their approach. Despite which route is pursued, managing the process provides security assurance and value. One solution is to deploy effective strategies and reduce on-demand requests. Another and probably the most valuable is to integrate forward and responsive thinking. Both of these address planning and reducing on-demand practices. The primary benefit is that forward-thinking utilizes predictive analytics and environmental knowledge, while responsive thinking focuses on ad-hoc, fixation, or unforeseen events.

Two drivers that model forward and responsive thinking is proactive and reactive strategies. In the context of cybersecurity, these are planning methods relating to security events or tasks. Although operational teams cannot predict the future, they can develop preemptive steps in reducing risks. Alternatively, when preventative measures are excluded, the environment becomes reactive. Both strategies serve and protect enterprises and require a balanced risk approach. It's somewhat similar to a balance scale—you cannot level the plate unless both sides carry equal weight. Similarly, the proactive and reactive strategies must impact the same features and operations.

Proactive security is a preventive strategy that "bakes" security into every baseline, architect, and service model. The methodology utilizes every security ingredient or holistic defense model to enhance security. It requires a security professional to think and promote defense while assessing, evaluating, and testing controls, policies, and settings, which models baking a cake. When the right ingredients are mixed, the outcome for the cake will be positive. Similarly, cake theory follows proactive measures. You start by gathering the security ingredients and formulating the mixture and best practices. The outcome mirrors the security culture and engagement strategy. Combining "bad security" into the mix will cause security gaps, additional risks, or distract cyber hygiene. Throughout every security steward's career, they have experienced some form of "bad ingredients." Although they may not have directly or indirectly "mixed the ingredients," the outcome was transparent—security failed.

The significance of proactive security stretches into many security areas. For instance, the assessments and evaluations are designed to discover vulnerable findings. A mindset that thinks proactively will onboard risk-based thinking and learn risk-reduction methodologies. When the attitude offboards risks, the environment may experience data breaches, attacks, and public distrust. The outcome produces the same results: "Where did we go wrong?" In 2019, the US Department of Health and Human Services experienced a data breach that affected twenty-seven million individuals. The breach root-cause never went public, but the impact was no secret—such as the security ingredients were "badly" mixed. When you establish security controls, you must also bake in security practices. So, the term "where did we go wrong" now becomes a part of the mixture. Sometimes it's best to mix the ingredients by following the proper practices, such as proactive security measures. Similarly, the mix for the ingredients includes situational awareness because our cyber senses promote bold decisions. As with the CDC and many businesses, a data breach emphasizes a security lapse, so performing baselines, assessments, audits, and defense tactics can counter data breaches.

Countering breaches starts long before data usage through configuration management practices. These are simple procedures that vendors or corporate security outline before networking or bringing services online. In the IT industry, businesses purchase many applications, infrastructure, or services from vendors. Each vendor provides a product best practice or configuration guide that states "change the default password." Despite the guidelines, numerous environments never change the default passwords; and when assessments occur, it is discovered as a critical risk. When deploying a risk-mentality, it becomes natural that being preemptive also means thinking outside the box. When configuration requirements are warranted, it should be a realistic practice to change passwords. This is often the most straightforward and overlooked risk. When referencing the inclusive culture, the procedure could be resolved through establishing norms. These are embedded working habits that speak risk reduction or driving proactive security measures as an enabler.

Business enablement is an integrated practice that combines proactive security and its initiatives. Each employee thinks, lives, and breathes security for every active engagement or task. Normal activities consist of a culture that speaks proactive languages. Picture a heavily engaged corporation—the entire enterprise, customers, and image would demonstrate frontline security. Alternatively, when business enablement has failed, the environment must operate from the rear and "catch up," which is where security fails the most.

Whenever the proactive security measures fail, the compensation leans toward reactive security. In this operational mode, the enterprise is fighting backward to counter attacks, risks, or critical incidents. The entire organization may have to operate under pressure, and many services and access rights are restricted. Imagine the availability of services and disruptions that occur. Let's not forget about the customers! So, with reactive security measures, the fixation and response procedures are a true friend.

Most organizations prepare for emergencies and incident response procedures. The reactive security approach is used to counter existing threats and vulnerabilities, which reduces risks. Teams spend a vast amount of time combatting damages that either hackers or system failures induce. To add, many enterprises practice incident response planning, but the real-time experience is far more beneficial. It provides open access to discovering the enterprise risk conditions and proactive security measures that either failed or need placement. Of course, the chances to gain real-time incident response experience are rare, so incident response training is valuable. The exercise allows IT teams to simulate and respond to various attack scenarios and create improvements.

When appropriately used, reactive security does have its benefits and rewards. The outcome provides risk state, program deficiencies, evaluations, and readiness indicators. Also, when cyber hygiene fails, it can drive and support points where weakness exists. This is where seizing the opportunity for success becomes essential— as it indicates what particular risk or practices are failing. Within any program, the strategy helps to sharpen security engagements and redefine proactive security measures. It's somewhat similar to a tune-up. At specific mileages or maintenance periods, a car will need recalibration. The parts may fail benchmark checks, so performing regular tune-ups can prolong the car's lifespan. Security operates in the same capacity for the practices, processes, or programs that require refinements. After certain intervals or system failures, each must undergo improvement due to reactive security. Specific benchmarks may indicate that each has internal issues or weaknesses. The improvements align with the growth mindset as the system is fine-tuned and works within a newer or advanced mode. The legacy design and security constraints or gaps are lessened or removed.

There is a mutual benefit for both measures—they share interdependencies and similar requirements. When reactive security executes, the success is demonstrated through preemptive engagements, which means the team must think ahead and organize the IRP. It's unwise to construct the IRP during or after an incident since risk reduction is the end-state. Alternatively, proactive security has a dependence on reactive events—it defines where your changes, refinements, processes, programs, or practices carry value. Some of its failing events may be categorized as newer or enhanced features when baselining. The term baselining indicates the minimum security controls and configuration requirements are implemented. When building a security architect, the failures are used to redesign or redefine risk treatment, which becomes a forward-thinking skillset. For instance, the outcome for assessments determines whether controls are adequate or deficient. The results further indicate whether the baselines are in compliance and what weakness exists. The information gathered is then used as a continuous improvement strategy. When the improvements are implemented, they are then classified as enhancements or refinements.

While performing many Risk Management Framework (RMF) assessments, enhancements and refinements became the norm. The RMF is a six-step process that integrates improvement and structured assessments for organization-wide security. As defined in the Computer Security Resource Center (CRSC), "The Risk Management Framework (RMF), presented in NIST SP 800-37, provides a disciplined and structured process that integrates information security and risk management activities into the system development lifecycle." Government agencies use the RMF to assess their security readiness and ensure systems operate safely through a Security Assessment and Authorization (SA&A) program. The entire (SA&A) procedure consisted of 250 controls and various configuration requirements. Our team used the SA&A to test each agency's controls and baseline compliance. After each test, the results indicated controls that complied and also which failed. During the out briefs, our team would provide best practices concerning control adjustments and configuration settings. Afterward, the agency would insert the corrections and create lessons learned. All the collected information was transformed into action items, which were monthly taskings. The action items would describe control checks and configuration standards that needed attention or verification. The follow-up teams would review the data and insert best practices, configuration changes, and enhancements for their perspective systems. The outcome lowered risks and prepared the agencies to surpass future SA&A cycles. In relationship to both measures, the agencies combatted risks early and practiced cyber hygiene. The typical failures were nonexistence, which resulted from forward-thinking and projecting a defensive mindset.

The defensive thinking transitions into security and provides the leverage to incorporate fixation and remediation exercises. It has been widely used within vulnerability management programs.

During its operation, the business becomes defensive when resolving patch issues, vulnerability remediations, or varying security changes. It's considered as an after-the-fact situation, which means something has negatively occurred, so let's defend the enterprise. In most technology platforms, the staff spends more time protecting security when protecting or defense security is required. The existence is due to regular security and criticality because many IT environments concentrate on external attacks. Although it's great to think defensively and protect enterprises, the offensive security practice is much more beneficial and valuable—it stops "bad security" at the gate!

Offensive security is where the "bad guys" are discovered through proactive engagements. Its goals are to discover preexisting deficiencies by performing manual and automated testing, such as penetration, web-based, vulnerability scanning, or audits. It's similar to any sports team that has an offensive strategy. Before a play is executed, the offensive team plans their movement. Their approach is to seek out the weakness in the opposing team and score. The flaw is discovered by analyzing their patterns, history, and defense strategies.

Similarly, the technology uses analytics, patterns, and events to prevent attacks from occurring. Each security team member represents a skilled player. Their role is to protect the enterprise and catch the "bad guy." They gain success using tools, practices, and staying proactively engaged. Each time the "bad guy" is prevented from entry into the system, the team scores. When offensive security activates, the hackers are hunted, and breaches and intrusions are prevented.

The defensive security approach reacts to the plays. In the same context, a defensive team has skilled players. Each player operates in the defensive mode: prevent the offense from scoring. As a defensive player, you must react to patterns and events and, hopefully, become the "bad guy" or "spoiler." In the technology space, the defensive security team is on standby and strategizing their roles or assignments. When disaster strikes, they must transition into a reactive mindset, defend the enterprise, or perform in emergency mode. Hopefully, their skillsets are tight, masterful, and disguised from the offense! Once the issues are discovered, there could be a confused state in how the results should be handled. After serving on standby and thinking reactively, its obvious vulnerable events will be found—so who is responsible for deciding the responsibilities and ownership standard? Many organizations are strapped and very afraid to face risk, so they continually spread the question around. Okay, it's our time to discover the answer—so let's investigate "Responsible Action and Ownership."

CHAPTER FOURTEEN

RESPONSIBLE ACTIONS AND OWNERSHIP MODEL

As the IT landscape evolves and onboards various security technologies, management must become task-based leaders. Every IT organization has projects assigned where leaders create tasks and drive expectations, and this is where task-based leadership unfolds. Many security professions are given work, and some assume them because of the corporate buy-in structure. Alternatively, other groups need a little motivation and direction. Despite which strategy is used, the projects require responsibility and ownership. Although not commonly discussed within technology, the phrase is a silent process. When security events surface, the situational awareness principles, proactive engagement, and culture inclusion model execute; and this is where responsibility and ownership are best demonstrated.

While onboard the USS *Spruance*, I experienced "Responsible Action and Ownership" and how task-based leadership operates. One morning, the division had a new sailor check onboard; and I will reference the sailor as "Hard Charger" since his uniform was sharply fitted, neat, and clean. To introduce himself, Hard Charger gathered with the division and presented himself with a unique glamor: "I am here to take over and no excuses." It was noticeable that he carried a rare and unique military approach. After greeting the entire division, he sat down since everyone was involved with maintenance checks. As the leading petty officer, I introduced my role and the whole division's. My role served to direct, supervise, and guide work management. Specific taskings include ensuring work was scheduled, completed, and reported for the ship's eight o'clock reports. The eight o'clock reports were nightly briefings that summarized the ship's readiness condition and maintenance status.

About twenty minutes into greeting and talking to Hard Charger, I received an email stating that the commanding officer needed the system's status report and its readiness condition. I transitioned my workload to the ships' Combat Information Center (CIC) to assist with the reports. The CIC was a designated location where the ship conducted information warfare, processed combat data, executed tactical engagements, and served as its battlespace. When entering the CIC, I saw Hard Charger in action and working on the critical maintenance task. What was noticeable was that Hard Charger was wearing his dress blues, and the uniform was generally reserved for watches or ceremonies. Also, he was using every available tool and maintenance guide for the task. A senior personnel named Master Chief X tapped me on the shoulder and said, "In my thirty years of serving, I have never seen a sailor take action in his dress blues." In my mind, this was the first sign of ownership! He was onboard for twenty-plus minutes and assumed responsibility and ownership.

When referencing the Hard Charger initiative, I also reference my post-sea-duty career as a naval leadership facilitator. For the assignment, I was charged with delivering leadership and development training to junior sailors. Subjects ranging from conflict management, teamwork, motivational strategies, organizational leadership, and career progression were discussed. Also, the students gained valuable information concerning how their actions influenced working goals and the Navy's mission. As a facilitator, I often referenced Hard Charger's accomplishments to drive many discussions. His level of ownership and taking responsibility was noticeable. Today, the memories of Hard Charger are everlasting and guide in building the risk mentality.

The leadership demonstrated by Hard Charger transforms into the cybersecurity sector. Each manager, supervisor, or technical leader encounters situations where responsible actions and ownership (RAO) must be demonstrated. Many risks indeed occur, but the control and responsibility toward a risk-based approach are internal. Varying practices from proactive engagements to cultural influence integrates into the methodology. When the concept becomes active, managers and staff personnel lead the best security and risk mentality culture. Their relationships and outcome provide a clear path into risk management best practices.

The risk-based approach requires every role or group to practice responsible security. Along the battle lines and planning strategies, it's common for CISOs to assume control and guard the enterprise. The rhetoric is more role-based than work-related, as CISOs hold a premier position and take responsibility for the corporate cybersecurity strategy. There is also a counter-argument that the experts hold the security keys and responsibility, which is partially true. Responsibility follows a top-down approach and also requires lower-level involvement. This does not indicate that CISOs are provided free passes in removing ownership. However, it does stipulate sharing responsibilities and individual rights are distributed. After all, the most junior security steward operates and uses the model.

When referencing the inclusive culture, the model drives the corporate environment. Every employee, team, or group engages with security and operates the outcome. It's a common practice where involvement exists, and ownership is demonstrated through various tasks and thought processes. When it is engaged, the corporation can influence the culture to accept responsibility and risk ownership. Alternatively, some employees may fail and shift blame.

Often, shifting the blame is utilized as an excuse when security fails; it widens security gaps and causes project failures. Security is about resolving the most complicated issues, so adding additional risks is harmful. By default, teams are very accustomed to shifting the blame. When engaging a structured program onboards responsible ownership, technical groups can resolve issues early, close security gaps, and remove blame.

The RAO model follows traditional leadership practices that define responsibility and accountability. The conventional security practices have had challenges and risk-induced practices where both failed. From my professional experience, I have seen security teams battle bureaucracy and politics and overlook a simple factor: "Let's build a responsible culture that can take action and assume ownership." To function as a responsible culture requires professionals to hold themselves accountable, take charge, and understand their role or group is a risk-conscious culture. All too often, the terms "responsibility" and "accountability" serve as mixed roles in cybersecurity. Responsibility is something that is shared amongst different technical groups. Within cybersecurity, there are various roles that employees assume to support risk management initiatives. They may function as a vulnerability analyst, security engineers, threat hunters, forensic analysis, or client support assignments. Each role functions to protect the enterprise and risk initiatives. The project manager may assign tasks according to role assignments and skillsets and mixed or confused responsibilities may exist if they are not articulated. The outcome creates additional risks and open actions—an area where risk remediation is overlooked. Also, people may never express ownership. Typically, responses such as "I was never told" or "it's not my responsibility" become a typical response.

When accountability exists, corporations can influence responsible ownership and individual actions. Team members realize they are held liable and cannot share ownership, and varying roles operate proactively and support risk remediation. When risks occur, each person knows their functional role and "takes charge." It's very similar to proactive security since the decisions, actions, and the risk-remediation outcome is self-owned, and when they "take charge" as a preventive measure in reducing risks, internally, they demonstrate a mindset that refuses to shift blame and also holds themselves liable.

In the cybersecurity culture, accountability is very mixed. Through bureaucracy and politics, paper-driven roles exist, but actual working roles differ. These are standard practices that decide the difference in whether risk-based approaches operate and whether risk-mentality exists. Each security function, group, or team must answer and support risk management initiatives. Answering is defined as "owning up to" the outcome. When audits or risk assessments fail, someone must be held accountable. All too often, it's not the case since ownership is considered foreign practice. Alternatively, the audit or assessment success still requires accountability since someone must answer.

Adhering to RAO has a direct extension into the "Code of Ethics." In the cybersecurity community, practicing protection is a defaulted responsibility. The cyber professional's role assignments come with added trust. The outcome enables thinking that ultimately protects the confidentiality, integrity, and availability of information assets. Key decision-makers, engineers, operators, and varying teams demonstrate behavioral norms and cultural-based involvement where risk-conscious decisions, actions, and engagements exist. Additional actions such as information sharing mirrors and technical groups operate in a transparency mode when weakness occurs. The entire concept models the situation awareness SAAF model in that "what we share" has a liable effect when ethics are violated, and risk decisions can either induce or reduce exposure. The result assumes that the initiator or group will respond with a take-charge mentality when induced risks occur. Alternatively, the unethical practice will shift blame instead of demonstrating ownership.

The Code of Ethics is mutually agreed upon in many security functions, policies, and requirements. One function that requires ethical agreements is cybersecurity certifications. Each candidate that passes certification examinations, such as the Certified Information System Security Professional (CISSP), Certified Ethical Hacker (CEH), or Global Information Assurance Certification (GIAC), must attest to ethical standards, privacy, and due diligence authority. When building the risk mentality, the certifications teach candidates how information protection and assuming responsible action and ownership reduces security weakness or induced risks. So, the question is how much-certified personnel follows the code of ethics. Considering that defaulted responsibilities and ownership exist more often these days, the majority would represent ethical practices.

The cultural influence and its image contribute to ethical behavior. As project teams develop, behavioral norms are practiced through task assignments. Most people understand that reworking a project is labor-intensive and stressful. So once the project norm executes, and ethical security unfolds, risk responsibility automates, and ownership becomes second nature. Also, it takes long hours and dedicated study time to pass a certification, so not practicing responsible ownership is a disaster. Let's be honest: responsibility and ownership exist well beyond technology and influence careers. Considering the amount of time, energy, and fees associated with IT certifications, it's an individual disservice and career risk when ethics are violated. Also, when vendors inactivate a certification, it directly affects employment availability and job hiring. As discussed, the code of ethics has more gain when properly followed vs. not followed.

One key term that's commonly used throughout the model is influence. The term influence suggests risk-based thinking affects security stewards' actions. When risk thoughts, concerns, or events occur, teams, groups, and varying roles must pursue relatable or assigned activities. The pursued action is based upon the affected environment, which can be internally or externally. Diagram 5 provides a high-level indication and an affected environment when the model is required or addressed. Each area assumes the methodology communicates a risk-conscious atmosphere and operates toward safeguarding enterprise resources, data, and information assets. Let's explore the four areas and analyze how RAO exists and formulates the risk mentality.

The first communication link deals with industry and business relationships. These are external agencies or corporations that depend on risk management decisions and ethical practices. When examining the relationship, it's evident that external partnerships establish a trusted interest in the internal risk culture and responsibility. They do not control or know the risk treatment programs. So with a trusted relationship, they assume that partners are ethical and risk-conscious. For instance, Corporation A has an audit that resulted in five critical vulnerability findings. Their impact could create a 50 percent reduction in system availability. Each industry partner carries the exact system requirements, infrastructure layout, service line, and risk model.

Let's take the concept further by creating a fictitious company name, "Corporation A." When sharing advisory information, Corporation A decides to report three vs. five critical vulnerability findings. In the past, Corporation A has been very ethical and followed the agreed reporting and sharing standards. The mission partner accepts the shared information, and within seventy-two hours, their entire network availability decreases. Corporation A states they shared the correct information. Now, who is responsible and accountable for the information shared? Corporation A's unethical risk practice has caused a severe failure. The future relationship becomes untrustworthy, and now, the partnering company becomes a risk-induced organization. Additionally, the term "ownership" does not exist since Corporation A demonstrated unethical business practices, and by default, they "shifted blame."

Many corporations share dependencies for risk management and information protection. Any company can be graded for risk responsibility through internal information, which means that internal assessments, audits, and security management structure identifies the risk approach. If correctly practiced, it leads to a more functional and secure culture that achieves goals.

When corporations draft their strategic risk plans, the entire organization is aligned toward goal achievement. The different team approaches and RAO are expressed and executed. Most organizations depend on the GRC program, while others set action items to address risk remediation or mitigation. The information gathered from the Tier I and Tier II teams is used to evaluate the security goals and progress. So accurate and detailed information becomes the difference between achieving or failing business goals. The methodology is quite different from false positives—as they are inaccurate results. False reporting is a deceptive and irresponsible decision that misleads corporate groups that the security posture is sound and compliant. All too often, teams will crunch numbers that sound very "green." RAO is about full transparency through reporting and communication. When establishing benchmarks, budget reports, and metrics, they also bridge the business requirements and cybersecurity goals. There is a trusted interest in knowing that employees, managers, and technical teams take charge when security fails. In the "Risk Discovery and Opportunities" section, a similar discussion was presented where discovered risks are modeled as positive results. There is a direct relationship and opportunity for success when ownership exists when examining responsible action. It provides goal achievement, full transparency for the security posture, and a risk-based approach that is goal-driven. Achieving such milestones requires a culture that onboards risk reduction. If we examine Corporation A unethical decision and business goals, the outcome will demonstrate the same result: irresponsibility and ownership refusal. Since the risk-based approach is designed to reduce weaknesses, the described actions would defeat cybersecurity best practices. How do you think this will affect the customer base?

Every technology professional and business end personnel is required to deliver customer service. Whether it's incident response events, service requests, or information sharing, each holds responsibility for corporate-to-customer service succeeding or failing. Customers are labeled as end-users, partnering companies, non-technical teams, board members, or new employees. The security service offered may involve data protection, account creation, website availability, or email communication. Every service offering delineates responsibility from various engineers, analysts, and operators and comes with ownership tags. If a responsible person untags their ownership, it differentiates between satisfied customers or submitting complaints. Imagine being a new employee and all the challenges associated with account creation. You may wait days before officially gaining system and email access. In your mind, you are thinking about whether someone cares; and when you finally gain access and the privileges are misconfigured, now you have the option of submitting a survey. Will you be bold enough to highlight the challenges? Welcome to customer success!

When invoking RAO, customer success means the detailed process and roadmap in resolving incidents, events, or risk findings are proactively handled, and customers achieve best practices. Each service offering—such as risk management—is regarded as a product. It's a relationship where businesses understand the security magnitude and a customer-focused mindset. The overall customer health rating directly aligns with the cybersecurity health practices, all driven through the take-charge attitude and assuming 100 percent ownership. Through positive and negative security outcomes, business lines attach responsibility and the right to sustain cyber hygiene. Each customer receives the best-in-class products and uses its features to reduce enterprise risks. The success is evident when clients use cybersecurity services. Each product owner assumes full responsibility and ensures that the customer demands and risk management initiatives are satisfied. If they are not met, the owner takes responsibility for correcting the issue—a true example of how proactive security and customer success operates.

There is a familiar term used throughout technology that delineates the responsibility boundaries and ownership guidelines called liability. When corporations describe the liable actions, they merely pronounce that providers, users, and clients operate within upstream and downstream liabilities. The cloud architect is one example of the upstream liability with the cloud service provider, and downstream liability is with the users. The liable actions are controlled through service-level agreements (SLAs), which dictate responsibility and data ownership. Typical SLAs would delineate responsible steps between or from groups, teams, business units, or departments.

Different groups establish upstream and downstream relationships. As a security steward, one has a dependence when various groups and teams are involved. The units can arrange from networking to the database, operations, or service support to provide transparency and communicate risk discovery and responsible ownership. The communication channel may require upstream relationships. For example, Team B has to share vital and accurate information with Team A and downstream information with Team C. Since Team B is the mediator, they must take ownership of data transactions and information sharing between both groups.

When deep-diving into the team and group information sharing responsibilities, various units such as vulnerability management (VM), threat intelligence (TI), and malware defense (MD) surfaces. Each team has a unique and responsible security function. The VM team must ensure that vulnerable findings are accurately reporting and assume ownership for the results. It means that they have to manage the vulnerability lifecycle and make every effort to remediate the vulnerability. If they remove ownership, risks and threats will escalate. Now, IT depends on tools, people, and processes that operate. When devices are out of compliance, the IT reports may contain inaccurate data. If the data is accurate, the IT program must enact data quality management (DQM). The purpose of DQM is to cleanse data reports and submit actionable and accurate information. This is where false-positive analysis becomes vital. Can we state that malware exists when our VM, TI, or DQM fails? DQM requires very accurate data, and when it's not correct, who will assume control?

In reference to responsibility, the issue can be remediated when responsible actions are targeted, communicated, and demonstrated. A security steward's role is to assume and manage responsible activity. Although one may not have direct involvement or decision authority, they must understand how DQM interfaces upstream and downstream responsibilities. With the number of ransomware attacks surfacing, the RAO model is well needed. Its core is about enacting proactive security measures to sustain MD and various security requirements.

The financial sector has also used the model for its shareholders' accountability. Each portfolio manager is required to see through financial decisions by taking complete ownership. The action alleviates any disconnects and confusing responsibilities. If there is a financial error, the account manager assumes responsibility and takes the appropriate action. Another way to describe accountability is "the act of seeing through a complete task." When referencing the security requirements, accountability aligns with risk management. It allows one to manage risk from an end-to-end perspective. That means once a risk is discovered, they take control and follow through with the remediation procedure. High-tempo actions may impede the remediation process in a realistic environment, so thinking proactive security would warrant a better solution. Again, the terms responsibility and accountability are owned by specific individuals, and shifting its course could cause the "ship to sink." If a security steward practices the methodology, the risk-based approach can become successful. Now, let's move forward and assess the risk-based approach! See you in the next chapter.

CHAPTER FIFTEEN

ASSESSING RISK MANAGEMENT PROGRAMS

Every organization has had some interaction and engagement for risk management programs. The architectural design, corporate functions, dependencies, and interdependencies dictate the overall program operational characteristics. Along those lines, they also incorporate expectations and stakeholder responsibilities and the corporate risk-based appetite. Many online discussions describe a risk management program, and by far, they all address a unified process—the program utilizes process, integration, culture, and infrastructure as crucial elements. Within each cycle, risk identification, access, measure, evaluations, mitigations, and monitoring are formalized and used to carry out the program. Some organizations have adopted variations of the concept, while others created similar operating philosophies. Despite which route is used, they all require a risk-based approach.

There are assessments within every risk management program. Their purpose is to identify weaknesses and security gaps and support risk prioritization strategies. Each organization that onboards risk assessment programs gains insight and proactive information about its security state. As stated under *Extending a Compliance Mentality*, the holistic defense model aligns to risk assessments. When evaluating an organization, the program must incorporate all dependencies, frameworks, concepts, and business practices. Various methodologies such as HIPAA, HITRUST, ISO 27005, NIST 800-30, or OCTAVE describe a risk assessment process, outcomes, and inclusions. Each method carries similar process steps and outlines the risk-assessment strategy. Depending on the specific organization and technologies, one or multiple techniques are used. It's common to see many organizations use NIST 800-30 since it's a government standard and well structured and customizable across varying platforms, systems, and organizations.

The NIST 800-30 operates in the assessment mode and utilizes NIST 800-53 control boundaries to determine its security. The government heavily depends on the NIST 800-53 since it's the "control reference;" it's the official guide supporting the DoD and federal agencies' risk management programs. Within NIST, 800-53 users will find seventeen control families and control implementation details. Each family addresses a specific security area and aligns to standard risk management programs. The benefit for NIST 800-53 provides a holistic control framework, so organizational initiatives are addressed.

It's important to note that the entire NIST Special Publications (SP) guides support ISO assessments. As stated in SP 800-30, "The concepts and principles associated with the risk assessment processes and approaches contained in this publication are intended to be similar to and consistent with the processes and approaches described in International Organization for Standardization (ISO) and International Electrotechnical Commission (IEC) standards. Extending the concepts and principles of these international standards for the federal government and its contractors and promoting the reuse of risk assessment results reduces the burden on organizations that must conform to ISO/IEC and NIST standards." Below is a summary of the NIST SPs and their overall relationship to the assessment process.

Special Publication	Description
800-30	Guide for Conducting Risk Assessments
800-37	Guide for Applying the Risk Management Framework to Federal Information Systems: A Security Life Cycle Approach
800-39	Managing Information Security Risk: Organization, Mission, and Information System View
800-53	Recommended Security Controls for Federal Information Systems and Organizations
800-53A	Guide for Assessing the Security Controls in Federal Information Systems and Organizations: Building Effective Security Assessment Plans

TABLE 16. NIST SPECIAL PUBLICATIONS

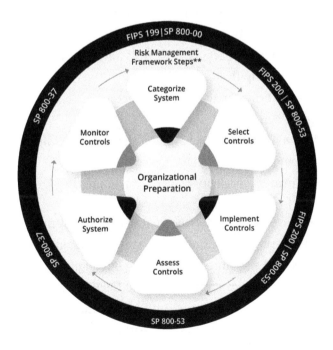

DIAGRAM-5: NIST RISK MANAGEMENT FRAMEWORK

Another addition to the NIST publications is the Risk Management Framework (RMF). The RMF is a holistic and comprehensive risk management process that utilizes security controls to satisfy organizational-wide information security programs. It also provides a method that integrates security and risk management activities into the system development lifecycle. Each iterative phase approaches risk functions and implementation standards that mature through the RMF process. It further supports organized initiatives by applying a risk-based approach to security control selection.

RISK MANAGEMENT FRAMEWORK PROCESS AND METHODOLOGY

1. **Categorize** according to the impact and a loss of confidentiality, integrity, and availability.
2. **Select** an initial set of baseline security controls based on the security categorization.
3. **Implement** the security controls and describe how they are employed.
4. **Assess** the security controls to determine their effectiveness.
5. **Authorize** operation based on the organization, procedures, and assets.
6. **Monitor** the security controls on an ongoing basis.

The RMF is used extensively during government security assessments, testing, and audits. Every agency is mandated to use the RMF process, and each integrates varying but aligned requirements. The Veterans Administration, Homeland Security, and Health and Human Services all use the program and add individual needs that align with the RMF policies. As stated earlier, the RMF is flexible and well-structured.

When referencing risk-based thinking, the RMF process can evaluate multiple technologies and its programs. It's a traditional practice to use risk management in assessing administrative, technical, or operational controls. An additional benefit exists when corporations use the RMF to evaluate their program. The entire methodology uses the RMF and defined Enterprise Risk Management (ERM) controls to promote risk decisions and objectives. Also, it addresses various risk-based thinking concepts such as proactive security, responsibility lanes, risk discovery, or combines the theories into an existing control framework. The outcome demonstrates the effectiveness of the ERM program and the weakness that exist.

For the remaining discussion, the focus will address assessing an ERM program. The objective is not to build readers into experts but to understand how the programs are evaluated. There are additional frameworks and assessment strategies that may work. In this discussion, the NIST methodology is used since it is well structured and allows flexibility. Under standard NIST RMF assessments, all six phases apply; and for the discussion, non-applicable stages will be omitted.

RELEVANT RISK MANAGEMENT FRAMEWORK PROCESS AND METHODOLOGY

1. **Categorize.** This will be used for learning, and it's not relevant to the assessment.
2. **Select** an initial set of baseline security controls based on the security categorization.
3. **Implement** the security controls and describe how they are employed.
4. **Assess** the security controls to determine their effectiveness.
5. **Authorize.** Not relevant to the assessment.
6. **Monitor** the security controls on an ongoing basis.

To demonstrate RMF use in a traditional risk-based thinking methodology, we will use the RMF methodologies and exclude relevant steps that do not apply. Since we will assess the ERM program, defining system categorization is not applicable. However, the process will be demonstrated for learning purposes. Usually, a system would receive a categorization rating after completing its security impact analysis (SIA). A unique feature is that it provides potential security concerns, security posture impact, and overall security risk level. Also, it evaluates baseline configuration, privacy, data sensitivity, and system categorization relating to Low, Moderate, or High controls. Table 17 provides an example of rated areas when conducting an SIA. Each relevant area is combined to form a "High Water Mark (HWM)." The HWM approach intends to ensure the appropriate security controls are applied and the required Confidentiality, Integrity, and Availability (CIA) exists. Webster Dictionary defines the HWM as "the time when something is most active, successful, etc." In this case, it's when security functions and the highest security safeguards are implemented. The following publications can be referenced for the categorization process.

- **NIST Special Publication 800-60 Volume I**: *Guide for Mapping Types of Information and Information Systems to Security Categories.*
- **NIST Special Publication 800-60 Volume II**: *Appendices to Guide for Mapping Types of Information and Information Systems to Security Categories.*

Information Types	Confidentiality	Integrity	Availability
Information & Technology Management			
System Development	Low	Moderate	Low
Lifecycle/Change Management	Low	Moderate	Low
System Maintenance	Low	Moderate	Low

IT Infrastructure Maintenance	Low	Low	Low
Record Retention	Low	Low	Low
Information Management	Low	Moderate	Low
System and Network Monitoring	Moderate	Moderate	Low
Information Sharing	N/A	N/A	N/A
Environmental Management			
Environmental Monitoring/ Forecasting	Low	Moderate	Low
Environmental Remediation	Moderate	Low	Low
Pollution Prevention and Control	Low	Low	Low
Economic Development			
Business and Industry Development	Low	Low	Low
Intellectual Property Protection	Low	Low	Low
Financial Sector Oversight	Moderate	Low	Low
Industry Sector Income Stabilization	Moderate	Low	Low

TABLE 17. INFORMATION TYPES

Let's assume that we wanted to choose system development, lifecycle/change management, system maintenance, and IT infrastructure maintenance as our information types. The initial step requires a CIA evaluation to determine the control baseline, which is the minimum safeguards to sustain protection. The outcome provides a control framework that identifies appropriate security controls and their functions. **Table 18** includes a list of information types. A rule of thumb is that the highest severity defines the watermark—so one "high" creates a high watermark for the CIA area. Here is the formula that represents **Table 18.**

1. **Confidentiality (L/L/L/L) = L, Integrity (M,M,M,L) = M, Availability (L,L,L,L) = L**
2. **Confidentiality (L), Integrity (M), Availability (L) = Moderate**
3. **High Water Mark = Moderate Control Baseline**

Information Types	Confidentiality	Integrity	Availability
Information & Technology Management			
System Development	Low	Moderate	Low
Lifecycle/Change Management	Low	Moderate	Low
System Maintenance	Low	Moderate	Low
IT Infrastructure Maintenance	Low	Low	Low

TABLE 18. INFORMATION TYPES—WATERMARK

As previously stated, the impact assessment is not relevant to the discussion. It's used as a learning vehicle for the RMF process. The government systems are categorized based on information type, so for this exercise, L, M, and H categorizations are excluded. This stems from assessing PM controls, which does not require categorization.

The next stage will onboard security controls and determine which control applies to the risk-based thinking. The *NIST Special Publication 800-53 (Rev. 5)* provides a list of security controls. The publication is primarily used amongst federal agencies and industry partners to execute an effective RMF process. For this discussion, the "Program Management—PM" controls were extracted and tailored. The areas identified in **Table 19** describe the applicable control families and their relationship to the risk-based thinking model. The option to use more controls would still suffice and also is dependent on the risk environment and business goals. For the discussion, the selected "PM" controls are designed for learning.

Number	Control	Description	Risk-Based Thinking
PM-1	Information Security Program Plan	Develops and disseminates an organization-wide information security program plan.	1. Extending Compliance Mentality 2. Risk Discovery and Opportunities 3. Proactive and Reactive Measures 4. Responsible Action and Ownership
PM-6	Information Security Measures of Performance	Develops, monitors, and reports on the results of information security measures of performance.	1. Extending Compliance Mentality 2. Risk Discovery and Opportunities 3. Proactive and Reactive Measures
PM-9	Risk Management Strategy	Develops a comprehensive strategy to manage risk to organizational operations and assets, individuals, other organizations.	1. Extending Compliance Mentality 2. Risk Discovery and Opportunities 3. Proactive and Reactive Measures 4. Responsible Action and Ownership

PM-14	Testing, Training, and Monitoring	The organizations provide oversight for the security testing, training, and monitoring activities conducted organization-wide.	1. Extending Compliance Mentality 2. Risk Discovery and Opportunities 3. Proactive and Reactive Measures

TABLE 19. CONTROL SELECTION

One of the additional tasks associated with control selection is tailoring. The logic behind tailoring allows organizations to select appropriate controls. For instance, an organization's risk-based structure may not onboard PM-14 since it is integrated under other control families. If the case were valid, PM-14 would be excluded from the control selection. The exclusion does not cause any significant gaps; instead, it allows flexibility in choosing what controls apply beyond the NIST recommendation. For our exercise, we will assume that each PM control in Table 19 applies.

Now, the control selection has been achieved. The next phase involves implementing the controls. The goal of the implementation stage is to ensure the rules operate and function within the risk-based model. We would want each control to serve toward risk management and ensure the controls align with security architecture. In our example, the PM controls align with the risk-based model, which is the architect. Another feature brought upon through the implementation stage is using a system security plan (SSP). The SSP serves as a living document that describes the operational environment, how the security requirements are implemented, and the relationships with or connections to other systems. Since we are accessing a PM family of controls, our interest is operational and security implementation. **Table 20** provides the implementation details for the PM family of controls.

Number	Control	Description	Implementation
PM-1	Information Security Program Plan	Develops and disseminates an organization-wide information security program plan.	The organization has implemented PM1 by ensuring a compliance mentality extends into a risk-based culture. Each risk discovered is addressed as an opportunity for risk success. The plan is updated on an annual basis, and it describes the overall risk management plan.
PM-6	Information Security Measures of Performance	Develops, monitors, and reports on the results of information security measures of performance.	Organization ABC utilizes KRIs and KPIs to measure its risk priorities and security staff approach to security defense. The overall proactive and reactive measures are used to determine gap analysis and risk-based approach.
PM-9	Risk Management Strategy	Develops a comprehensive strategy to manage risk to organizational operations and assets, individuals, other organizations.	Organization ABC has implemented a risk management strategy that outlines how individuals, teams, and management approach security. The process further explains how the organization's security posture should operate through normal and abnormal security conditions.
PM-14	Testing, Training, and Monitoring	The organizations provide oversight for the security testing, training, and monitoring activities conducted organization-wide.	The organization's training and readiness program onboards a stringent and operational training cycle. The organization focuses on training that aligns with the mission objectives and security requirements.

TABLE 20. CONTROL IMPLEMENTATION CHART

The next phase focuses on testing the control effectiveness and operational requirements. Each PM control will be assessed and provided an outcome that describes its effectiveness. Risk discovery and opportunities are used to grow the risk-based program and view the program beyond compliance. The assessment strategy operates by reviewing, examining, or testing selected security controls, as described in **Table 20**. The process utilizes teams, completed tasks, or events that represent the risk-based process. Also, the assessment strategy evaluates the PM controls against the implementation description, which helps identify weaknesses and security objectives that require attention. **Table 21** provides an example of the assessment strategy.

Number	Control	Test Description	Test Outcome
PM-1	Information Security Program Plan	Verify the security plan is operational.	The tester used events that required organization ABC to use its security program. Each use for the plan provided sufficient support to conduct the following: 1. Assess the likelihood and potential damage of threats, taking into consideration the sensitivity of the personal information. 2. Evaluate the sufficiency of existing policies, procedures, customer information systems, and other safeguards in place to control risks.
PM-6	Information Security Measures of Performance	Validate there exist performance indicators such as KPIs and KRIs that are being utilized.	The organization's previous risk management program was evaluated to determine its effectiveness. Samples were extracted from two metrics that defined how the measures of performance were successful. 1. Information system security personnel that have received security training showed was rated at 95%. 2. Systems in compliance with organizationally mandated configuration guidance produced a KRI of 90%. The score was above the standard.

PM-9	Risk Management Strategy	Validate the ERM strategy onboard PII, privacy, and risk-based concepts.	A comprehensive test was conducted to determine whether an ERM strategy exists. Samples were executed from previous risk assessments to verify whether specific risk-based initiatives exist. The tester evaluated the following. 1. The privacy program was used to reduce data breaches and sensitive exposure for PII. 2. Each exposed risk was monitored and inserted into a Corrective Action Plan (CAP).
PM-14	Testing, Training, and Monitoring	Validate that the ERM program addresses testing, training, and monitoring.	The organizational annual Security Awareness and Risk Monitoring was implemented. Every year, the results are documented and tracked. 1. The organization tested the employee's response to phishing attacks 2. The Security Awareness training was completed by 98% of the employees and monitored through quarterly reviews.

TABLE 21. SECURITY ASSESSMENT REPORT

The last objective focuses on monitoring the security controls. It will use the assessment-provided indications and status for the security controls. The Information Security Continuous Monitoring (ISCM) phase considers the security control status, operational requirements, and risk-based process. The entire monitoring process is designed to align the security requirements and controls so the risk-based approach can function. Each CAP is addressed and monitored for remediation. Throughout the entire risk-based lifecycle, the compliance mentality, risk discovery, and responsible security practices are onboarded.

Overall, the Information Security Continuous Monitoring (ISCM) furnishes ongoing observation, assessment, analysis, and diagnosis of the cybersecurity posture, hygiene, and operational readiness to support organizational decisions. It enables organizations to transition from a compliance-based to a risk-based mindset. It provides organizations with information necessary to support risk response decisions, security status information, and ongoing insight into security control effectiveness. The following areas describe the functional components of the ISCM.

- **Continuous Assessment**: Supports ongoing assessment of security control effectiveness, reporting and monitoring security risks.
- **Continuous Reporting**: Increases situational awareness and supports reporting requirements and trends in overall residual risk, broken down by inherited risk, accepted risk, and risk to be mitigated by CAPs.

Using the NIST RMF process is a valuable tool for assessing risks. As stated before, the process is well-structured and provides flexibility toward ERM strategies. The NIST RMF is evolving as the cybersecurity industry matures and onboards different security objectives. The corporate arena and federal agencies must change their strategy as well. The changes help to articulate security from within vs. without. When security functions operate internally, enterprises can build a more resilient architect and create proactive defensive measures. The outward effect or external causes await responses. The goal is to operate along the front line and proactively combat each threat-induced situation. Through risk discovery and risk-based thinking, the cyberwar can operate ahead of hackers and protect enterprises. Achieving such an objective requires cybersecurity professionals and organizations to transform their mindsets—the methodology onboards every security discussion through virtual thinking and risk-based concepts. We are in a digital age where transformational thinking can improve cybersecurity. Now, let's go ahead and "Transform the Mindset."

• TRANSFORMING THE MINDSET •

- Building A Value Proposition Mentality
- Thinking Digital Modernization
- Modernizing A Workforce
- Wearing A Hackers Hat
- Adaptative Mindset

In July 2019, the Department of Defense (DoD) publicized its Digital Modernization Strategy for FY19-23. The guide addressed many technical challenges from artificial intelligence to cloud innovation. Its strategy was driven by different threat landscapes and technology growth that accelerated past the government's current architect. Many of the DoD components utilized legacy technologies and processes that were immature or risk-prone. The priorities such as cloud posed significant changes where the DoD had to modernize its services and technologies to align and adapt to industry standards. The core for the cloud, artificial intelligence, cybersecurity, and Command, Control, and Communications (C3) relied upon modernizing and evolving the current practice into a transformational framework.

Transforming the current technology architect for corporations is a very comprehensive and well-thought task. The core requires people, processes, procedures, and decisions to operate cohesively, which means modernization onboards their growth. Yes, the entire scope is about development. In relationship to the growth mindset, the concept aligns the cybersecurity culture to cleanse the legacy tactics and modernize security. Typical engagements require cloud technologies that operate within the customer mindset and security framework. Legacy processes defined security as a separate entity and lessen room for innovation. Many corporations and DoD components operated under the concept and never onboarded a modernization plan that transformed their mindset, security engagement, or growth potential. So not having a modernization plan would position the DoD to operate within a risk-prone environment. Alternatively, the end state—when used correctly—optimizes security services and risk-based approaches. Technical teams gain confidence knowing that newer technologies provide a modernized avenue to protect systems and operate ahead of hackers.

The word "transforming" is defined as a change in structure. Some examples are reorganizing a security program to remove older technologies or costly processes. When aligning the definition of the Cybersecurity Mindset, it discusses new and innovative strategies. The current security state is positioned to support more recent technologies, innovation, and cost-reduction methodologies and perform vital roles that will shift the cybersecurity perspective into a streamlined environment. Within the environment, the time, labor, and costs are structured to optimize cybersecurity. One particular group that utilizes the concept is technical teams. Each team uses a framework that transitions security into an operative environment. Their security mentality onboards data protection and IT modernization standards. Typical outcomes exhibit more straightforward methods to manage security taskings and resources. The conventional approach may operate through ad-hoc procedures that cause overlaps, which eventually disrupts time management and decision making. Just one failing function can cause a chain reaction— so this is why transformational thinking is a successful practice.

When transformational thinking surfaces, organizations can foresee rapid improvement and cybersecurity maturity. The frontlines of security are mentally armed and positioned to defend enterprise assets and reduce risks. The success is by far dependent on the transformational methodologies used to grow the security environment. Environments may consider removing or restructuring teams, changing workloads, or outsourcing projects. Another method is to centralize cybersecurity, which removes individual teams, processes, and service lines. Through a decentralizing methodology, the transformation involves separate and functional security practices. Although this seems risk-prone, some corporations have survived well in using the concept. Despite which method is used, the transformational mindset must onboard a structured process that's successful. One program that works well is the Value Proposition Mentality because it builds a starting point for understanding the transformational mindset and its benefits.

CHAPTER SIXTEEN

VALUE PROPOSITION MENTALITY

In 2019, various corporations and federal agencies onboarded growth strategies. The change escalated from attacks, phishing attempts, and data breaches that compromised their systems and applications. The concept associated with benefits, buy-in structure, and capability maturity models guided their decisions and success factors, and every transformation initiative absorbed security best practices. During the phase, each migrated strategies to align customer needs and military operations—because they desired services that offered value-added changes and cybersecurity reshapement. The entire transformation provided agencies an opportunity to think about internal and external valuation and why developing business-to-business (B2B) relationships is crucial. Key stakeholders and internal agencies shifted their mindset and identified value-added services that would ultimately mature their posture. As a result, the agencies developed relationships and navigated cybersecurity with success; and the number of remediated attacks created value. The outcome and reshapement integrated a Value Proposition Mentality (VPM) where new offerings, business integration, and risk reduction successfully operated.

The VPM is a reshaping tool that allows corporations and federal agencies to think and build best-in-class security products. The concept operates through a multi-channel environment where the customer and product owner succeeds. It's common for security services to support customers, but through a successful VPM, the customer and business gain rewards. When security value allows customers to defend their technology platforms, reduce risks, and mature their cybersecurity posture, the provider can establish upward and downstream business relationships. Their partnering companies (upstream) can gain an edge in knowing their security service evolves, and their customer-base (downstream) security architect becomes well protected. As each technology platform matures and assumes more needed assistance, the previous relationship surfaces. Both remember the initial benefits and success gained from the first B2B relationships. Their mentality focuses on "value" and what was achieved. The end-state establishes the VPM as a vital contributor and enabler that changes security thinking into a modernized environment.

When attempting to incorporate the VPM, the methodology must be communicated amongst staff and technical teams, as they are the core enablers that drive the VPM success. Managers may face challenges and struggle to pitch the value-added service, which results from communication gaps. Some typical offerings are protection models or staff realignment. The protection models are enhanced procedures that transition the security response and risk management programs into newer frameworks. The enhanced concepts may require the incident response plan (IRP) to be reconstructed based on recent attacks and containment strategies, which is time-consuming and labor-intensive. The entire documentation library must be updated, and every attack scenario must be tested via a table-top exercise (TTX).

A TTX is designed to test the security response procedures through emergency services, which means the organization is in combat mode. Each scenario is based on real-time operational issues and events, and lesson-learned exercises created to enhance the TTX. So trying to sell the value-added position for the TTX is a challenge. One strategy that supports the situation is to engage security teams before the change. The engagement consists of communicating the security posture and defensive mindset early and often. As the managers share the defense status, the staff can foresee their value-added position in action. If newer technologies are onboarded, the team will establish an automated buy-in structure and think about innovative ways to support the transformation.

As a manager, you have successfully communicated the protection model; now, the entire organization needs to be realigned. Trying to realign the security staff and sell the VPM is a demanding task. Every security steward has a comfort zone; it's that place where role familiarity and routine operations provide job security. When they are tasked to perform security objectives and anticipate the outcomes, their position becomes stagnant. Now imagine an attempt to realign their role.

Organizations may cause confusion and create resistance. If leaders communicate the value, the security staff can onboard transformational thinking and optimize the corporate security requirements. This eventually leads to understanding how IT security services are modernized and how their outcomes create lean and scalable programs, secure operations, and strong defensive tactics. These optimized programs will eventually require less time and different skillsets to defend the security architect. As a security steward, one would be thinking about their benefit: "When the security landscape transforms, then I will use fewer resources, modernized skillsets, and gain efficient processes to defend against attacks." The thought process is very significant, as it provides early entry into security relationships that will support cybersecurity practices. Each security professional and the business unit can integrate standardized and modernized services and ultimately eliminate unnecessary resources, applications, and systems. The outcome drives businesses to minimize their threat surface and vulnerabilities. Imagine a security steward who understands a staff realignment is a value-shaping tool vs. additional work. A corporation could gain success and value recognition across multiple security engagements and partnerships.

As a manager, you have built success around the communication channels. So now, let's talk about prioritizing your value-added service. In a typical security environment, there exist many offerings and business capabilities. As a company begins to mature and onboard capabilities, it can provide a more diverse service. These capabilities are defined as security programs and operational support such as threat defense, auditing, malware prevention, or vulnerability management. The capabilities improve cybersecurity for personally identifiable information (PII) and all actors who use cyberspace to transact business and service offerings. Ongoing cyber threats, risk initiatives, or defense tactics can also operate and provide B2B and customer confidence. When these objectives are met, the value proposition becomes more defined and streamlined. All too often, companies inaccurately define their capabilities and priorities beyond their qualifications. The process leads to elevated risks for customers. It's a challenge for the security environment to manage normal risks—so inducing more risks can defeat security best practices and cost reduction strategies. Also, an inaccurate capability profile affects the ability to reshape cybersecurity. Imagine a company attempting to reshape a customer's security landscape, and its value-added service is misaligned. Can the customer goals be met? How will the B2B relationships operate? These are pondering questions to answer when offerings are inaccurately defined through business relations.

One viable solution that supports accurate capabilities and service offerings are streamlined management. It describes how corporations can optimize and prioritize their business model. In the context of streamlined management, corporations dissect both and identify their customer requirements. Both can be as simple as delivering more affordable and effective cyber risk practices that support the customer. The government contracting program operates along the same lines through its statement of work (SOW). The SOW is a document that defines the government requirements when products are needed. Each submitting contracting company must identify their strengths and whether their capabilities support the government requirements. Contracting companies pose great benefits and chances of winning a contract when their qualifications align with the government requirements.

When transforming cybersecurity, streamline management can serve as a potent tool. It allows companies to align their resources and capabilities during contract proposals and to determine whether their business is security-focused. When delivering service offerings, a streamlined approach starts with the company and then the customer. Suppose the company streamlines its services and emulates a Cybersecurity Mindset. In that case, it can cultivate the best-in-class protection standards, defense tactics, and cost-reduction strategies. The path in gaining such achievement starts with the "Virtual Path"—Inclusive Culture, Situational Awareness, and Risk-Based Thinking. These are the core enablers that shape transformational thinking and simplifies the cybersecurity mentality.

Another area that uses streamlined management is service optimization—it navigates a company to align and define its value-added position and capabilities. Many companies defer using the practice and utilize individual practices to decide security goals. The concept is labeled as decentralization, and it operates independently from the enterprise security framework. Although the decentralization architect allows more flexibility and independent operations, it also breaks security across many entities and establishes various value streams. The security engagements may also work along different boundaries and produce varying security outcomes. The design defeats security as business goals are to demonstrate optimized services and capabilities that support value thinking. Controlling and prioritizing benefits can be challenging since communication lanes are broken, and each business unit operates separately.

A viable solution that reduces independent operations is a centralized security architect. In this model, corporations think about streamlining their services into an enterprise framework. Each outlining unit operates from the enterprise, and the corporation becomes much leaner and agile. They can effectively manage security and deploy defensive tactics that function through varying transformational stages. Typical security gaps imposed through changes are minimized since the enterprise control security operations. Leaders can control risk appropriately within the model, deliver aligned security services, and support value propositions and their prioritization. It also ensures capabilities are aligned and followed. Through the decentralized approach, we learned that each business unit independently thinks and articulates its capabilities. In a value-centric environment, the process operates much differently— each business unit utilizes the same capabilities. Also, the centralized system provides flexibility and ongoing observation into changes, which advances the value of safeguarding the enterprise.

In 2009, the US Navy utilized the centralized methodology to onboard security programs and operations due to the DoD transformation. The agency was directed to upgrade its systems and consolidate networks, applications, and downsize operations. Various changes occurred that required technology transformation and offboarding of legacy systems. The entire transition produced an environment where growth and maturity integrated into security best practices, operations, and value-centric relationships formalized. Each sub-component for the Navy received support in modernizing and driving efficient security defense strategies. For instance, various commands failed short in managing cybersecurity traffic and analyzing data feeds. Due to the security gap, incidents escalated, and threat response teams found themselves cleansing and remediating more incidents. When onboarding Computer Network

Defense (CND) programs, the incident activities were reduced drastically, and threat response teams could better analyze data feeds. Without integrating growth strategies, the US Navy would endure high-cost and additional risks for its IT systems. Thanks to the CND transition, it advanced the US Navy's ability to add value and integrate resilient IT systems.

The term advancing cybersecurity emulates growth and increases readiness. At its core, growing is about thinking, improving, and driving innovative protection standards. Every business or federal agency utilizes holistic defense strategies to outsmart hackers and conduct risk management programs. These skillsets combined with a VPM enhances service offerings and allows customers to protect their data, applications, and infrastructure devices. The US government is well versed in using the concept when contracts are awarded. Each SOW outlines various technical requirements and needs that will resolve problems. The SOW may specify that the agency wants to upgrade its networking security system and reduce forecasted incidents. As an offeror, the capabilities offered must align and match the requirement. The capabilities must demonstrate performance and value. Many times, offerors fail to outline their value proposition, and other offers are awarded the contract. Maturing the value-added solutions is the most viable answer for the issue, as it provides an outlet to support cybersecurity maturity.

When discussing maturity, it requires the organization to think, grow, and promote business capabilities. They align value-added service models that demonstrate the company's worth and also advance customers' cybersecurity goals. Both contribute to security readiness and advancement. To successfully operate the model starts with internal growth strategies. These strategies describe how companies can position their skillsets, capabilities, and services through marketing channels. Every company has a unique offering that supports customers' business goals. The offering is designed through business development milestones and performance areas.

As a company decides to structure its capabilities, they examine their skillsets, labor, and performance profile to determine how best to benefit customers. Since each customer carries varying requirements, the company must design, inventory, and onboard skillsets, qualifications, or certifications that demonstrate their worth. The skillsets reference cybersecurity performance such as incident response, threat hunting, malware prevention, or cloud services. As the company gains additional skills and matures, they become more marketable and a high-value asset (HVA) to customers, partnering companies, and investors.

The rapid access to the best labor also supports capabilities and service offerings. Each employee comes with many skillsets that demonstrate their contribution and benefit to the inclusive culture and the employer's business objectives. Suppose a corporation desires to win contracts supporting penetration testing or cloud security services. They must onboard top talent who possess high-level certifications and skillsets. Having a full range of training and skills means Company A can approach the most challenging security objectives. It's more defined as having an arsenal of cyberweapons, which is the combative effort that logically protects information assets, interfacing systems, and resources.

In the government contracting space, the best-in-labor references proposals and service offerings. Within the government regulations, various stipulations determine what skillsets and certifications are contractually related. As a solution provider, the talent acquired must model the federal contracting requirements. Having employees with significant certifications such as the CISSP, CEH, GIAC, or CCNA means your capability model has the best talent. Just think about the employee named "the Jack of All Trades"—the customers respect their value. The employer also benefits, because it highlights cyber strength and a strong VPM.

While starting SEMAIS in 2014, the same methodology surfaced when developing the business capabilities. There were multiple opportunities to onboard capabilities in every security category, but what value could SEMAIS offer? The entire capability statement was written about ten times before formalizing the draft. What was evident and learning was that the capabilities had to carry value and support the government contracting requirements. Each statement, differentiator, and profile description was driven through a customer's mindset, as they wanted to strengthen cybersecurity. How did SEMAIS contribute? The firm positioned its mindset, capabilities, and service offerings as an HVA. This meant our service modernized customers' security architects and matured cybersecurity.

The final delivery for security transformation is to think about HVAs. These are the services that reduce risks and allow customers to channel growth. Thinking about HVAs is a mental state that drives corporations to manage persistent cyber threats and enhance business relationships. As services are delivered, and the value proposition expands, the B2B relationship becomes solid. So do you think the customer would change vendors? Probably not since the B2B relationship and HVAs work as a single entity. The modernization climate reduces cost, secures data access, improves defenses against malicious cyberspace activity, and controls risks. Business relationships become valuable and less expensive over time. So when the entity knows its capabilities and delivers a high-value service, it becomes an HVA. A long-term B2B relationship can also build security familiarity, which means the customer or business security architect is well understood and protected.

In today's market, transforming the mindset seems overwhelming. Different technologies and cybersecurity services support various interoperability standards. They carry unique requirements, so deciding which helps the cybersecurity architect is a burden. For instance, when should AWS and Azure be used or considered? Also, what modernized technology can deliver security protection and reduce vulnerable actions when multiple platforms exist? These are questions to ponder when corporations desire growth and transformational thinking within their technology space. Also, resilient systems and securing technology becomes a factor.

Organizations require resilient systems to support unforeseen incidents, attacks, and intrusions. To leverage the requirements, designated sources must prevent, detect, and react without considerable loss and drive continuous availability. This is where the technology space and transformational thinking operate. Their outcome promote technology communities to "think digital modernization" and accelerate cybersecurity across multiple technology platforms.

CHAPTER SEVENTEEN

THINKING DIGITAL MODERNIZATION

In 2014, the technology market was slowly expanding and designing modernization strategies to advance cybersecurity. The cloud technologies were ever-increasing, and technology integration for many other services such as IoT, web-based applications, artificial intelligence, and automation developed its service capabilities for commercial and federal entities. As a business owner and security enthusiast, it became apparent that the trend was starting, and innovative ideas would surface into cybersecurity. When attending many government conferences such as TechNet, which was sponsored through the Armed Forces Communication Electronics Association (AFCEA), the technology innovations and showcases were interesting. There existed over three hundred vendors and government-affiliated companies that displayed innovative products. Here is where modernization and cybersecurity surfaced, because information sharing, networking, and technology demonstrations afforded a rare opportunity to experience technology transformation. Where else could a display of technology innovation focus cybersecurity thinking and how innovation is the key to modernizing defensive strategies?

Many speakers and military personnel discussed how the Army desired to be leaner, automated, and advanced in their cyber defense programs during the conference. One speaker provided a very straightforward session about the Army's reshaping and transformational thinking. The speaker indicated that the Army was onboarding additional requirements to support the warfighter and their cyber operations through innovative ideas and automated processes. He also irritated that the outcome was to ensure their cyber warfare programs support the Army's mission: create a cyber ecosystem that would defend the Army's information, data, and infrastructure domain. After the session ended, the notes were reviewed, and the thoughts, actions, and thinking toward digital modernization and defending ecosystems became clearer. What was needed from a business end was a digital modernization capability model and service delivery framework. With both, the driven effort in combating security and growth challenges would succeed; and today, it's apparent that the trip to TechNet created an early entry into thinking digital modernization.

The simplest definition of defining digital modernization is through innovation. Corporations and federal agencies realize that security outcomes place challenges into defending their landscape or ecosystems. They continually solicit contracts, hire additional labor, or onboard newer technologies to pace with the security needs. Sometimes the outcome produces a very defensive architect and business model, and then there are times when the innovations occur every year. Innovation is about reforming and reshaping technology into a more modern framework. It positions entities to align their architect, services, or security environment to business partnerships or customer requirements. For instance, software applications were once loaded locally onto computers and servers, but they are cloud- or web-based today. The modernization service allows corporations to concentrate on operations and transition maintenance to their cloud service provider (CSP) through the cloud. The web-based applications provide easier access across various departments and regions and promote a centralized management practice. When locally operated, the maintenance was executed on each device, which is time-consuming and expensive.

Another functional area that describes modernization is evolving. The term outlines how technology molds itself into industry requirements. As security incidents occur and expand, corporations are thinking about protection needs. Their incident response program (IRP) may need to expand and onboard different response procedures that stem from newer attacks, phishing, or malware. This is quite different from innovation since an IRP involves technology growth and changes. Through innovation, newer services would exist. It's somewhat a design or IRP practice that never existed. When accurately used, evolving brings security thinking into perspective and addresses program, operations, or business security structure.

When starting the security journey, most of the attacks were network-based as hackers were penetrating infrastructure devices— and many of the common attacks include Denial of Service, Man in the Middle (MIM), SQL injection, or Spoofing, which was where network security and business concentrated. Over the years, and as hackers became sophisticated in their attacks, software applications became their next target. The .Net applications became vulnerable, and database programs became a hacker's appetite. In applications such as SQL and Oracle, millions of records contained private and financial information, and the outcome drove the need to protect Personal Identifiable Information (PII) and Electronic Public Health Records (ePHI). Today, PII and ePHI have become critical security flaws, and everyone thinks privacy—it has become a legal obligation. Law firms, hospitals, insurance agencies, and learning institutions think and embrace privacy.

When examining IT history, it shows that thinking modernization advanced the technology space to become more innovative. If modernization did not exist, where would technology operate? The entire landscape would be vulnerable and allow hackers to expand their appetite. Having a Cybersecurity Mindset enables the culture to think about modernization. They understand that the evolution process has driven society to utilize advanced techniques, technology, and programs, continually evolving since protecting information, data, and applications is ever-changing. Whether it's the corporate arena or the federal space, the methodology works and provides the entrance into understanding digital modernization.

The transition into digital modernization is a procedure that requires extensive thought. It's apparent that sophisticated attacks, monetary loss, and customer-based service are forefront, but additional motivators are digitizing cybersecurity. These motivators stem from investment concerns that are outdated or pose significant security risks. Every organization's business process model evolves and requires some fine-tuning. Sometimes developing a new and innovative procedure can be less expensive and serve as an industry alignment. For instance, when infrastructure environments incorporated varying devices, they inherit various attack vectors. These attack vectors provide entry points or a gateway into the corporate network. As the machines are added, vulnerabilities increase, and data breaches' potential becomes a high-level risk. When there are three hundred users in an IT environment versus three thousand, the risk profile rises significantly due to foundational requirements and entry points. Having more users requires additional technology and security—the attack surface increases due to different software, hardware, and web-based applications. Imagine a corporation's digital footprint expanding to 500K users. The entire environment would need varying technologies and a digital mindset that constantly performs. Each week there would be different issues and security concerns that drive management to think, *Do we need to modernize our infrastructure and operations or define newer process improvement models?*

The typical phrase "Why should businesses modernize enterprise IT?" is a thought process that drives modernization. To assume the transition entails monetary costs and technology innovation, for a business to examine whether it's beneficial or not is an intelligent decision. Once the "modernization train" starts, the organization must commit costs and technology risks. Project managers can best explain the modernization train, as many companies start the process, then decide to regress. What's their reason? Well, it's probably due to inadequate research or changes within their monetary budget. Despite which happens, security thought and digital protection needs is a strategic decision. Early in the strategic process, CISOs must drive the thinking model and assemble research teams that understand the corporate security posture, operations, and landscape. Hiring outside consultants can very well work, but corporations still need to have internal involvement. The internal teams can develop a more comprehensive research proposal that models their experience and the security culture.

During the COVID-19 pandemic, the Department of Veterans Affairs (VA) transitioned to a digital modernization platform. The effort supported medical care, human integration, and remote working technologies. Many workers required remote services and software programs to manage 21.8 million health records tasks and telemedicine requirements. Various contracts were awarded, and typical in-house service transitioned to online and mobile technologies. So, how did the VA succeed? It's apparent that they were seated on the modernization train, but the pandemic accelerated their need. Before the pandemic, the digital transformation train had already shaped the VA's vision, as the organization's IT spending increased between FY2020 ($2.4B) and FY2021 ($3.9B). The $1.5B increase shows that the investment into digital modernization was a well-thought process, and COVID-19 provided security concerns that required more investments.

Today, the VA is making additional investments and taking many strides to modernized onboard technologies such as Electronic Health Record System (EHR), which stores health information and tracks patient care. The system connects the VA medical community with the Department of Defense, the US Coast Guard, and private medical facilities. The EHR program provides information sharing of veteran health records and accelerates data exchange when emergencies occur. The traditional system required veterans and medical facilities to share health information manually. The EHR program eliminates the burden through electronic and centralized-sharing processes. Every medical partner will view a single electronic health record and can update and securely exchange patient data. Imagine a veteran innovating paper-driven methods to share health information—a significant data breach or risk is posed. So what's the value in enabling increased information sharing for the medical partners? First, the EHR program optimizes medical record access and centralizes healthcare data. Logging into several systems may retrieve erroneous or missing health information because of the database connectivity or privileged access. Second, the program reduces informational risks. When several medical providers use different systems that are decentralized, the information risk is heightened. A centralized system provides single-system access and a data repository.

The EHR program contributed to removing legacy technologies. Its outcome reshaped the VA's modernization efforts and secured single access. In the modern area, IT organizations utilize legacy systems due to budget constraints and operational commitments. When these systems were built, they presumably used older operating systems (OS), and upgrading the system required the removal of legacy applications. Depending on the application criticality, the OS may need to remain. Microsoft-based systems face more issues with legacy operations. The Windows XP, Windows 7, and older .Net applications are no longer supported, but many organizations still use the systems. Although it's a risk, the organizations may have rated the risk as low or are awaiting additional budget. If these systems are internal, then the threat is lowered, as no information is shared outside the environment.

Thinking digital modernization and legacy systems is an innovative practice, but there are trade-offs and significant risks. Removing a critical application is a risk within itself, and using the legacy OS is a risk. So, what are the options? Well, most organizations modernize by transitioning the application requirements to an updated platform. For instance, when data is stored on older databases, the information could be transitioned to SQL or Oracle applications. Of course, there must exist successful data portability and interoperability requirements. Portability states that the database can be exported to another program or environment and function. It's somewhat like moving an object from box "A" to box "B," and it fits into box B architecture. Interoperability suggests that the services and operations for the data will remain. Can you still retrieve the data and manipulate the fields and tables after moving the object from box "A" to box "B"? Does the table show matching the last and first names? Also, when using phone numbers, does the updated system contain dashes? If the updated design is not dash-driven, the data quality program needs to address data cleansing before migrating. These are good thoughts to assume when legacy database applications are transformed, removed, or reshaped to meet security requirements and digital modernization standards.

When transformational strategies are developed, the entire plan focuses on IT and innovative technologies. The plan's survival heavily depends upon the investment undertaking and organizational security culture. It sounds compelling and technically correct in onboarding technologies that support customers and B2B relations such as IoT, cloud, web-based, and Endpoint and VPN security. Underneath the entire process, it relies upon security concerns and protection needs. No organization can undertake an IT modernization practice without researching security integration. The newer technologies require extensive research and whitepaper analysis to foresee and categorize security weaknesses. All too often, the process is omitted from security best practices. When migration occurs, the organization experiences security obstacles and additional risks. The risks can be monetary, technical, or program-related. Now, the regression occurs, and the financial spending becomes a loss. The point does not indicate that modernization is a precarious transition. However, it does stipulate that organizations must consider security costs.

When thinking about digital modernization, the holistic defense model and risk-based awareness become its success factors. The entire landscape and technology interactions, support requirements, and induced risk factors must be researched and analyzed for trade-offs. The best management and research strategies consider industry-approved products and services. However, the transition into an organization architect may not be conducive because the services add more weight, and IT modernization without a holistic mindset is dangerous. The security business is about strategic planning and driving the most successful protection standards, so that holistic defense tactics can address total security. Eventually, the thought process leads to a more resilient security architect.

Bridging the holistic defense strategy into security proves that more resilient thinking is being practiced. Resiliency drives defense by ensuring that the enterprise security state sustains, and the corporation can "fight" during a degraded mode. The sustainment occurs when different events, changes, advanced persistent threats, or system disruptions occur. The business goals are still met through the resilient process while preserving its security state and operational requirements. When modernizing security, resilient thinking determines success or failure. As different systems, applications, infrastructure devices, and assets are integrated, the IT architect must withstand and recovers rapidly from disruptions. It's somewhat similar to a rubber band—no matter how far you stretch it, the rubber band reforms its shape. The concept applies to digital modernization as resources are removed or added and adverse conditions occur—the critical capabilities must continue operating. These capabilities are the core enablers that keep a business functioning. Typical capabilities are the usability and availability of a database that supplies user account information. Without the device, the organization's availability to manage customer accounts becomes degraded. Depending on the degraded level, the entity may have to halt service and operations. Now, will the customers be happy?

"Think Digital Modernization" is a practice that can be embedded into many security functions. The technology industry innovates various solutions to support security. As these technical solutions mature, an entity must integrate thought processes that represent its position. Although the business may not support IT modernization, it must continue learning, researching, and discovering its digital modernization position. When the industry matures, they may foresee avenues where digital modernization is required; and some critical areas are usually overlooked, such as services and workforce skillsets. One could argue that both are equally important, but skillset and labor alignment are far more critical. Businesses can purchase the best products and services and still have security gaps. Today, many organizations experience the issue, so what's the answer? It's maturing the holistic defense strategy and organization-wide risk management programs, which commonly address technology, not skilled talent. If businesses transition their thinking to build the best workforce solution into the holistic defense model and risk management functions, they could become more secure. The solution drives technology and talent to operate in parallel—and also stimulates workforce modernization. We want an organization that modernizes a workforce and innovates strategies that enhance roles, labor, and workforce development initiatives.

CHAPTER EIGHTEEN

MODERNIZING A WORKFORCE

The technology industry has had a vast number of integration and transformational events from 2020 to 2021. Before the pandemic existed, many collaboration programs were mainly used by IT and business lines. Since the crisis, educational, religious, sports, and news outlets relied heavily upon the tools. Programs such as Microsoft Teams, Zoom, Cisco WebEx, and GoToMeeting usage expanded. Newer workforce requirements drove the underlying change. For instance, the National Football League utilized the collaboration tools to hold the 2020 draft; and all thirty-two teams held team meetings and virtual practice sessions. Despite the change, the NFL was able to communicate everyday information to players and coaches effectively. Workforce changes and digital modernization marked complete success. The entire US workforce, including the NFL, transformed remote technology into a virtual landscape, meaning every workforce entity or culture became modernized. The traditional means of workplace communication slowly dissolved, and today, collaboration programs are the norm.

In 2008, the DoD cybersecurity program was taking many leaps and changes. The dire need to secure infrastructure, applications, and data became a high-priority and serious undertaking. Since the environment focused more on securing military tactical equipment, a discovery indicated that commercial off-shelf (COTS) products needed the same attention. In its transformation, the DoD positioned itself to modernize training and IT certifications. In 2005, the DoD integrated the 8570 instruction that outlined specific training and IT certifications. Before certifications existed, candidates received employment based on resumes. After implementing the instruction, the candidates had to demonstrate their security knowledge via certifications. The entire change revealed how the DoD modernization mindset included the cybersecurity workforce's future and their skillsets.

Modernizing a workforce is about ensuring the organization has aligned its entire labor and skillset requirements. At its core, the environment realigns many programs, labor categories, teams, and security objectives for its security needs. Older security practices for the culture may become obsolete due to technology integration and incidents. That's right! If too many attacks are occurring, the environment has to cultivate changes, as every organization's public confidence is critical. The concept has a direct alignment with risk-based thinking. If workforce modernization is very proactive, entities can drive security from the front. Of course, the reactive method is when an attack occurs and transformational thinking surfaces. We want to cultivate the workforce in the proactive stage, and this relies upon situational awareness. When gaining information concerning the environment, organizations can predict workforce-related risks and effectively modernize them.

One work-related risks are tool usage and availability. As technology develops and environments transform, the most up-to-date and proper tools are required to sustain visibility. Each tool should be assessed against dying skillsets and vendors' End of Lifecycle (EOL). These are extremely sensitive risk enablers. Typical devices such as Splunk, Nessus, QRadar, or IBM Big Fix provides actionable and vulnerable data concerning enterprise risks. The workforce will suffer if either tool fails or does not supply the performance to defend critical assets. Now, who do we blame? Regardless of responsibility, just remember that thinking ahead saves security!

The workforce modernization extends beyond tools and also addresses skillset development. When onboarding different talent and labor categories, each business unit must provide adequate and modernized training. That means specific cloud services require onsite or remote talent to manage their technology. Its planning requires a strategic initiative that supports digital modernization. What's the secret? They must cultivate holistic defense and extract workforce development requirements. It's successful in having the most advanced technology, but integrating the "right" talent provides a more secure and thriving environment.

When the technology boom occurred in the 1990s, there was a dire need to have computer programmers, scientists, and mainframe operators. Every major corporation hosted skillsets and technology coding such as C++, Pascal, FORTRAN, and hardware such as 28.8Kps modems. Each technology professional wanted to embrace Bill Gates and become the next programmer. As the technology industry matured and data became much more helpful, the skillsets slowly diminished due to new requirements in securing information. The IT workforce changed due to skillset risks and antiquated technologies being underutilized. During the same period, many military technologies transitioned from tactical to COTS-based operations. Over time, the labor risk and skillset became less valuable since technology professionals utilized newer programming languages such as C++ 17, ASP.NET, Python, Java. Since the older programming skillsets were less required, the workforce reshaped its technical requirements. Now, the risks began to surface. Older workers were somewhat at a disadvantage because of aged skillsets and technology innovation. For instance, the Common Business-Oriented Language (COBOL) is the oldest high-level programming language invented in 1959 for mainframe systems. The language was used in creating business applications, and since mainframe computers' use slowed, the skillsets and labor shortages surfaced. Many COBOL workers found themselves unemployed or forced to learn advanced languages. Today, the language still exists for older applications and has since been replaced with modern object-oriented programming languages. Now here is a twist! The COVID-19 pandemic proved that government systems had significant problems in processing data, as the current use of Java did not automate report formatting. So COBOL was still required since most state and federal systems needed a language that automatically formats data. What's the risk? When these COBOL workers retired, their knowledge followed; now, the industry is strapped to find a solution.

The COBOL concept applies to modernizing the cybersecurity workforce. Many changes occur when updating training and organizational alignment, as both require the best skillsets and talent. It goes beyond just discovering the "right talent" as companies must consistently mature their training programs and realign skillset requirements. For instance, as the cloud, data analytics, and mobile applications gain more use, business units have to realign their staff and assemble security teams that communicate, respond, and fulfill organized objectives. All too often, role assignments are mixed or confused. In some organizations, there exist overlap or omitted security objectives that induces additional risks. So what's the answer? Well, the use of skillset analysis can curve the risks. The concept requires workforce managers to evaluate skill sets and technology when transforming, reorganizing, changing, or downsizing its cybersecurity program. The procedure extends past organizational charts and emphasizes talent placement, which may require human resource support. The job ads, labor categories, pay structure, or training are modernized to support talent acquisition. With so many security titles and skillsets in demand, workforce modernization must also incorporate talent management.

The purpose of talent management is to attract, recruit, train, and manage workforce requirements. It's an ongoing effort that matures and provides rapid access to cybersecurity. Its maturity process follows the workforce growth and requirements. The program utilizes the best-in-class tools and strategies to ensure a capable workforce exists and can defend enterprise assets. Imagine trying to determine workforce needs with a paper-driven methodology. Could managers calculate trends accurately? How can skillset gaps be monitored? This is where modernization takes precedence—we must also align the tools and programs for the cyber community. There are many functional parts, but the skillset gap analysis component provides a more unified procedure to determine security workforce needs and objectives.

So, you have successfully organized and attracted the best talent! Now, the system is due for an upgrade. Well, the mindset states that training is required. In every organization, there exist products, skillsets, or certification training. Each has an independent enabler and requirement, so when workforce managers engage in training objectives, they perform deep analysis and research learning requirements. The analysis is drawn from incidents, risks, and security changes. If the entity experiences more attacks over a specified period, they could consider additional skillset training. It's so common for corporations to view technology integration as a single training enabler. Managers must incorporate programs, processes, and trend analysis data to determine training needs when modernizing security. The data obtained serves as a direct enabler to meeting digital modernization objectives to equip the growing workforce and cultivate the best talent.

Each year, various IT talent becomes challenged due to the technology growth. When the cloud surfaced, the IT industry was just thinking about defending onsite applications, data, and systems. The cloud brought upon different and similar security objectives where the skillsets became cloud-based. Various entities transitioned their architect and realized cloud-based skillsets were in demand. When thinking of modernization, the market requires strategization, as it's the gatekeeper to defending enterprises. The skillset gap analysis is drawn from many scenarios and data-gathering activities. You will never completely understand what skillsets are risk-prone unless an incident occurs or objectives are not met. The analysis examines labor categories and their job duties to determine whether they align and support security objectives. As entities transform, their skillset analysis should mirror. You cannot technically develop without maturing or transitioning aligned skillsets. Any business that fails the process will induce more risks. Our role is risk reduction, so thinking digital protection extends into many functional programs such as skillset analysis.

The cybersecurity community can drive skillset changes and reduce risk exposure. When performing a security gap analysis, the skillsets must be analyzed as well. They serve as proactive measures to prevent broken safeguards, align defense tactics, or drive protection standards. Imagine a new firewall whose configuration profile supports internal-facing or private communication. In its setup procedures, Network Guy configures the firewall for public-facing communication. What's the risk? The internal IP addresses become publicly available, and a hacker's appetite is fed! The entire risk profile has just increased, and Network Guy will be a non-Network Guy. If asked about the error, Network Guy may introduce, "I assumed." We can help Network Guy in many ways, such as determining what and where skills are required. The new firewall may have had a manual vs. automated setting that determines interface connections. If Network Guy was given training during the product evaluation, he could grasp this difference and think past assumptions. All too often, beliefs override skillsets and break defenses and proper operations. When new technologies are created, we must always think about security and skillsets. If one fails, so will the other one.

In 2013, the Department of Homeland Security (DHS) introduced the most comprehensive programs to address cybersecurity and skillset gap analysis. The National Initiative for Cybersecurity Careers and Studies (NICCS) program served as an online resource for career, education, and training information. As stated by the DHS in 2013, the program was brought forward to align the rapid demand for cybersecurity skillsets. The DHS believed that as technology advanced, the United States needed a capable workforce to protect the nation's vital infrastructure and assets and align skill sets to combat the ever-changing threat landscape. The entire transition proves that workforce modernization expands into many sectors and provides closure of the same resource: skillset.

The DHS furthered NICCS in 2017 by creating the National Initiative for Cybersecurity Education (NICE) framework. Many federal agencies needed a more formalized plan to determine their roles. Each agency had multiple skillsets and objectives that were overlapped and misaligned. The NICE methodology resolved the challenge through role assignments and providing a more defined workforce. It allowed employers to use focused and consistent language in professional development programs, industry certifications, academic credentials, and workforce training.

Skillset	Description
Collect and Operate	Provides specialized denial and deception operations and collection of cybersecurity information that may be used to develop intelligence.
Investigate	Investigates cybersecurity events or crimes related to information technology (IT) systems, networks, and digital evidence.
Operate and Maintain	Provides the support, administration, and maintenance necessary to ensure effective and efficient information technology (IT) system performance and security.
Oversee and Govern	Provides leadership, management, direction, or development, and advocacy to effectively conduct cybersecurity work.

Protect and Defend	Identifies, analyzes, and mitigates threats to internal information technology (IT) systems and networks.
Securely Provision	Conceptualizes, designs, procures, and builds secure information technology (IT) systems, responsible for aspects of a system and network development.

TABLE 22. NICE WORKFORCE FRAMEWORK

So now that NICE and NICCS have been introduced, how will they support the workforce modernization initiatives? As a training manager, human resource professional, or CISO, the program leverages a very structured format to identify skillset gaps and transform security into well-defined labor areas. When transformational thinking occurs, it's best to realize that much of the work has been done. Through research and studying the industry, managers can discover innovative workforce resolutions. Imagine a company building a framework similar to NICE. It would require a very labor-intensive working model and dedicated hours to study its outcome, where optimization is needed. Along with optimizing security services, the industry can modernize its processes. Using a pre-built and already-functioning workforce development program offers ease and flexibility and room to tailor individual needs. All too often, optimization focus on technology concerns and monetary expenditure, not expanding outside the box. Improving workforce development is about innovation and streamlining services so the arsenal of talent can succeed. Yes, there is an arsenal of talent, and they rely heavily upon skillsets, so let's position success.

There are many functional elements to workforce modernization. The areas discussed are just the high-level requirements that create success or failure. One could argue that individual accomplishments and employee incentive programs are also inclusive. By far, they would be correct, but the focus is to deliver high-level information so managers can successfully think about their decisions. Yes, management decisions significantly affect how the methodology operates. If they practiced 90 percent of the workforce modernization concepts, they could presumably balance skillsets and risks. Each employee would gain the most sophisticated skillsets and defend what matters most: data, privacy, and information. When the opposite occurs, the "hacker's appetite" becomes active and successful. We are not into feeding hackers! So, how do we become sophisticated in our approach? Well, wearing the "hacker's hat" seems more conducive and relatable. See you in the next chapter!

CHAPTER NINETEEN

WEARING A HACKER'S HAT

When the IT industry started to evolve, the biggest concern was having data availability and network connection. The end-users utilized modems, legacy computers, and simple applications. As the industry progressed, the need for data became a priority and heightened risk. Consumers began to gain access across multiple locations, and then the ease of data became important. To its existence, mobile technologies surfaced and provided easier data access.

Along with the data growth, many transactions began to occur since consumers shopped online. Also, the IT industry experienced an upward trend in data access points, and so did the protection and confidentiality requirements. Within the background, criminals lurked through private and financial data, and system intrusions became a critical concern. The intrusions were mainly data access and network exploration where hackers wanted financial information or specific data. Today, the same risks exist, and hackers use sophisticated tactics to gain access and invoke worldwide panic.

When we hear about intrusions, it's often too late since the harm is executed. For instance, when the Equifax data breached occurred in

2017, the data records affected 147 million consumers, and Equifax's public confidence was in question. The root cause revealed that the Equifax patch management program was ineffective and caused exploitation. Now, every hacker thinks and takes advantage of an environment that lapses security. Any defect, lapse, or unpatched software program can create a "gateway," as every intrusion needs an access point. Every day their appetite is satisfied, and some organizations contribute to their calling—"Let us infiltrate your system." An online site called Websitebuilder posted significant statistics that support the claim. The website addressed the growing cyberattacks and their nature by compiling key statistics concerning cybercrime, as outlined below.

THE TOP 10 CYBER SECURITY STATISTICS AND FACTS

- Seventy-nine percent of organizations were expected to be affected by cybercrime in 2019.
- A malicious hacking attack occurs every thirty-nine seconds.
- Routers and connected cameras account for 90 percent of attacked devices.
- Usernames such as "root" and "admin" must be avoided, as well as passwords that duplicate the username or are a variation of it.
- The Yahoo data breach in 2016 compromised three billion accounts.
- The average cost per lost record is $148.
- Eighty-eight percent of companies with over a million folders have at least 10 percent open to all employees.
- An average company protects just 3 percent of its folders.
- Fifty-eight percent of data breach victims are small businesses.

The information posted by Websitebuilder is crucial to understanding the hacker's environment. Hacking is far beyond network infiltration, as it can disrupt operations and endure financial loss. The actors or groups behind each hack know the desired outcome and carry similar thoughts and capabilities. They are defined as "the lone wolf," digital perpetrators, or "nerds." Despite the labeling scheme, the definition of hacking is simple: it involves the illegal intrusion or infiltration of systems, networks, applications, or information. The reason is quite simple as state actors work through their country's government. Then, there are local groups that operate within silos or cells. These hackers often look for personal gratification and satisfaction and digital power. Just one keystroke can cause world panic! They are constantly lurking and researching vulnerabilities and thinking about weaknesses, such as unsafe passwords, default accounts, unpatched systems, or software defects.

Hackers can be further categorized as ethical or non-ethical. Ethical hackers carry a mindset to prevent intrusions. They operate through third-party organizations or internal teams. Their goal is to validate, test, and penetrate the system. Businesses gain advanced information concerning exploited vulnerabilities, and they can become proactive in system defense. Unethical is straightforward. These are illegal intruders that digitally harm networks or systems. Their thoughts and perceptions are to cause havoc. The Equifax incident is a great example, proving that a hacker's mental state is highly damaging. So what contributes to their thought process or behavior? We can state power, but there also lurks a "criminal mindset." To defend against hackers beyond the contribution, the "hacker hat" is worn and exhibits the hackers' mentality and a criminal mindset. Typical criminals perform physical crimes and exploit vulnerable people. Hackers follow the same principle, except it's digitally orientated.

So, what's the answer to wearing the hacker's hat? The most concrete objective is to understand the hacker's environment and transform it into ethical practices. The entire unlawful practices can be analyzed and used as a security development approach. The digital modernization process requires a holistic defense model, and thinking with a hacker mindset is by far a strong contributor. The outcome provides an early indication of the security posture and vulnerable conditions. Remember that risk discovery and opportunities provide proactive defense measures, and it's a component of ethical hacking. The hacker's hat is not a physical concept; instead, it's the thoughts and actions that proactively defend enterprises. When it's successful, organizations can transform and safeguard critical resources. But first, they must integrate and utilize the hackers' environment, which holds various objectives such as penetration testing, listed vulnerabilities, tools, tactics, and techniques. Let's put on our hats and explore the mindset!

Let's assume we were in a cell with twenty digital criminals and discussing how we could infiltrate an online banking system. The entire discussion would start with information sharing. Here is where we would begin to examine specific tools, vulnerabilities, and targeted organizations. If the organization patched their environment frequently and remediated findings, our role would be more challenging—as the vulnerable software and hardware would not pose any risks. So as the group discovers, trying to infiltrate software programs would fail. Now, let's twist the concept around. Do normal IT people share information? Probably so, but not to the extent of hackers. Hackers operate in silos and cells hidden from the normal society, which means they are an inclusive culture. As the mindset transforms, the same should apply. We are an inclusive culture that prevents unlawful entry and shares hackers' information. It sounds like the Information Sharing for Situational Awareness (IFSFSA) model!

After all the information sharing, we finally discover a vulnerable access point. We are so happy and feel powerful. But first, we need to learn what tools to use. We realize every environment may not even have our tools, so we are ahead of the curve. A security manager's nightmare is about to escalate. The manager decides to research, and he finds that the tool XYZ is commonly used in system intrusions. Since he did his homework, he made it known that the tool is required. When performing ethical hacking, the goal is to mimic, mirror, and utilize a hacker's tool bag. All too often, the tool bag is removed, and risks escalate. Under the hat, there exists a defense strategy labeled "tool bag usage." Managers must articulate that defense strategies are an intruder's nightmare through digital transformation—so wear your hat!

Now it's time to intrude the system and gain more action. We have decided that system entry is achieved through new firewall access. The company had a recent concern and posted its IP address in a support queue. The partial information provided direct access to the firewall, and next, we begin exploiting the vulnerable access point. The concepts mirror penetration testing, which is designed to penetrate and exploit vulnerabilities. Most companies have never had a penetration test performed, which makes them vulnerable. A penetration test detects flaws early and provides risk indications. It mirrors the same scenarios and environment unethical hackers practice, such as their tools, tactics, and techniques. Some entities may never experience an actual penetration exercise due to the monetary costs. Although the financial situation may be expensive, the return on investment is far more beneficial. As a security manager, the ROI is your number one challenge since intrusions are costly. We were successful at intruding the firewall—so mission accomplished!

When replicating an intrusion, the current threats and attack scenarios must be pre-thought and included. The actors use them to intrude networks and create backdoors. Just think of an ethical hacking exercise that used outdated or legacy hacking scenarios. The training may focus on older applications that are not installed or rarely operated, so does it make sense to exploit the applications? The current threat environment and hacking tactics provide the best indicators. It's typical to test for SQL, buffer overflow, cross-site scripting, or denial-of-service attacks. However, a more successful program assesses the hackers' environment. This does not alleviate the mentioned attacks, but it does provide a very efficient due diligence process. Ethical hacking is a replicative process, as entities should fully mimic or mirror the hackers' environment.

The replication process also extends into strategic thinking and business goals. Since hackers operate in an ongoing and continuous environment, defense strategies should mirror the same practice. Through building, designing, and engineering systems, the hacker's hat should be continuously worn. All too often, the holistic defense principles exclude the minor elements, and enterprises become vulnerable. In a functional and operative holistic state, ethical hacking and system security are resources that protect enterprises. In the earlier chapters, the holistic defense model was described and articulated how security components resolve risks and continuously master protective chains. Risks elevate when one link is either broken or misused and the "hacker's appetite" is satisfied. All personnel defending applications, systems, data, and information should mirror the same thoughts and security actions. If entities think and breathe the unethical hacker mindset, they can become mentally and technology aligned and invoke security actions that state "protected." The security gaps and excuses are never mentioned since their active involvement works in a proactive state.

Driving defense starts with the system design principles and thinking tools, tactics, and techniques hackers utilize to infiltrate IT systems. We could well state that security controls are the best solution. However, it's more of examining the design principles and the nature of intrusions. How, why, when, and where intrusions occur dictate the engineering and protective state. For instance, when digitally modernizing hardware devices that require newer applications, the system design mentality must incorporate end-point risks and how intruders use the assets. The more assets organizations add, the more risks exist. Every gateway represents an entry point. As organizations add more assets, the number of entry points increases. Another topic that falls within this category is end-point security, which is the defensive schemes used to secure end-points (i.e., desktops, laptops, smartphones, tablets, servers, workstations, Internet of Things or IoT devices). We are in a mobile environment, so the end-points become more vulnerable, and the defense strategy must secure its "hacker mentality."

The discussion for endpoints is critical. Here, code execution attacks, open ports, and OS information's misconfiguration can leverage outside attacks. As the mobile industry grows, so will the inherent risks. Within their architect exist high-value assets that cybercriminals target. If intruded, it makes the business difference and security survival. How intruders gain inside information is simple: they target critical and high-value assets. These devices are either openly labeled or known by hackers through research and study. The infiltrators discover simple knowledge about traffic analysis, network-connected devices, and unmanaged systems in their research. Just think about a database that supports clients' financial information and access. The database's use, design, and operative aspects should remain private through system thinking. When system design principles work outside the idea, the risk of a data breach surfaces. It sounds like information-sharing supports the hackers' mentality. What, when, where, and how information is shared shapes the holistic defense model. Remember that the unethical hacking principles and system design model are inclusive; if not, potential risks can exist. Working in the opposite state can detect risks, which means thinking and breathing unethical hacking in a continuous mindset is very defensive.

Wearing the hacker's hat can be disregarded as a symbolic practice. It's more of a simple metaphor used to enact the hackers' mentality. The environmental nature of hackers is straightforward: let's intrude and cause panic. When operating digital transformation strategies, incorporating the hackers' mentality curves security gaps and increases system protection. The responsibility falls beyond just management decisions and requires security action. The actionable practices are people, processes, and procedures. People must think logically about intrusions and how they occur. The process is simple: how do hackers succeed, and how can we become advanced in defending against hackers? When the technology industry grasps the concept, they better plan and coordinate before intrusions occur; transforming the mindset matters most. Thinking, breathing, mirroring, and mimicking the hackers' mentality is a new thinking mode. We can virtually learn how security is shaped, but transforming the mindset means security personnel adapting the "cybersecurity" thoughts and required actions.

CYBERSECURITY MINDSET MODEL

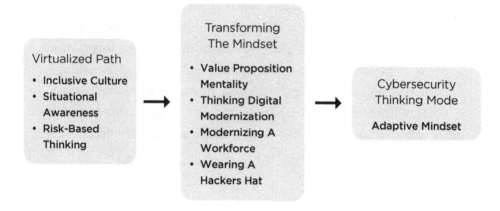

Virtualized Path
- Inclusive Culture
- Situational Awareness
- Risk-Based Thinking

Transforming The Mindset
- Value Proposition Mentality
- Thinking Digital Modernization
- Modernizing A Workforce
- Wearing A Hackers Hat

Cybersecurity Thinking Mode

Adaptive Mindset

CYBERSECURITY MINDSET: A VIRTUAL AND
TRANSFORMATIONAL THINKING MODE

• CYBERSECURITY THINKING MODE •

When building an arsenal of security best practices, they eventually become our toolkit and playbook. We use containment, strategies, and tactics to challenge the ever-increasing security attacks and operational modes. These environments are where security normalcy surfaces and designs a simplistic approach to staying ahead. Falling behind articulates that security is ad-hoc or has lost its structure. What the industry requires is a thinking environment where the cybersecurity mode of operation is active. When this occurs, the cybersecurity ecosystem projects the most comprehensive safeguards, watchful attention, and security best practices.

The "Cybersecurity Thinking Mode" is best described as the integration of the Virtual Path and Transforming the Mindset. It allows an enterprise to operate in a normal state of readiness. One can argue that readiness is a high-level status where security is good.

In some perspectives, it's true, but the readiness condition establishes functional cybersecurity practices. That means the good, bad, and average security is operating, and the enterprise is functional. The business can adjust and make operational changes that reform its security state, progress, or initiatives through various change cycles.

The Websters' dictionary has presented a formative definition for readiness: "the state of being ready or prepared, as for use or action." Without putting much thought into the description, it presents an exact alignment to readiness. The business or security-involved team must be in an active security state to deploy holistic defense. Yes, the entire environment must support actions and responses, which stems from situational awareness. An alternative explanation is "guarding the environment," where security teams are constantly available.

The availability of people and security builds operational readiness—the entire cycle onboards operational practices and procedures to challenge and reform normalcy. In any environment, it's expected for changes or events to occur. The reaction provides an accurate readiness indicator and where availability stands. The cybersecurity thinking mode brings all these principles to attention and drives holistic security. The entire picture involves thoughts, thinking, and a proactive mindset, preparing the business to adapt and remain resilient.

CHAPTER TWENTY

ADAPTIVE MINDSET

Cybersecurity encompasses many functionalities and protection standards that constantly evolve. Each functional element contains the technologies, processes, and policies that navigate, prevent, and reduce risk exposure for the cybersecurity ecosystem. Many experts acknowledge that the dire need to concentrate on end-users' attitudes, beliefs, and practices is warranted. The concept is very accurate by far, but change happens at a constant speed within technology's crevices. As technical experts gain an information advantage, develop counter-tactics, and improve defense strategies, the technologies evolve and require a transformational process. Each affected environment must reexamine its culture and branding to ensure its DNA, cyber hygiene, and security actions remain focused and aligned. Some organizations can successfully onboard changes without interrupting their cyber operations, risk mentality, or digital modernization plan. Alternatively, others fail and increase risks.

Cyberspace can produce many uncertainties and test an entity's risk exposure and defense strategies. One day the operational nature can succeed and defend the most hostile attacks. There are times when the attacks gain an advantage. Management teams are usually bothered and wonder, "Where did we go wrong?" Through forensics and investigative practices, the team may discover that their adaptive mindset was unsuccessful. Afterward, the technology team decides to develop adjustments. For the future, they envision a very productive and responsive ecosystem that remediates risks. Although it sounds convincing, there are intricate considerations to developing an adaptive mindset. At the core, there is a resilient and sustainable mentality. A "360-degree" of security visibility at the perimeter defeats intrusive events and safeguards critical assets, information, and applications. If either fails, the adaptive mindset will lapse and require adjustments.

The adaptive mindset comprises the Inclusive Culture, Situational Awareness, Risk-Based Thinking, and Transforming the Mindset into a constant and vigilant process. When practiced, many of the security challenges are rapidly mitigated or remediated without considerable interruption. Each business unit knows its security state and operational condition and can execute a secure and resilient landscape. Their intelligence gathering and situational awareness provide advance notifications and proactive response, which also enables availability. Another feature that describes an adaptive mindset is anticipating security behavior and preventing threat escalation, risks, or attacks. When utilized, companies can rapidly think and scale their services, operations, and defensive strategy upward or downward. The process requires a profound understanding of the current landscape and defensive patterns. A new employee can think and adapt in some capacity, and cultural challenges may exist since they are partially branded. Sounds familiar! The inclusive culture and adaptation carry similar relationships. When trying to adapt to changes, the first impact is the culture. You must understand the culture brand and drive resilience and sustain security for its operational requirements. What is ultimately achieved is a culture that logically adapts and integrates risk-based thinking.

Many IT projects utilize risk-based thinking and adaptive mindset principles, such as vulnerability programs. While working within the vulnerability management space, there have been many instances where adapting to technology and risk changes became challenging. Each month the enterprise would execute vulnerability scans targeting over 100K devices. The results were posted and analyzed to determine the patch and remediation plan. As the network, system administration, database, and desktop teams scroll through the two million data records, they prioritize remediation based on severities and vulnerability age. Some vulnerabilities were remediated, and then others were either missed or rated as a low priority. The cycle was continuous, and many teams questioned why the vulnerability count was aging, and some issues were never remediated. They failed to understand that a vulnerability management program is ever-changing and newer vulnerabilities will surface. We often state that management never thinks of adaptation since vulnerabilities outperform their knowledge and approach. Just imagine the pace where different vulnerabilities occur periodically, daily, monthly, or after each patch cycle. The irony is that the adaptive mindset must be tuned into knowledge and understanding priorities. Each month there is a chance that particular assets are not scanned; then, the following month, the scan occurs, increasing the vulnerability count. When the adaptive mindset operates, management can drive thoughts about operational challenges and adjust. Their adjustment may focus more on executing the priorities or designing innovative strategies to reduce the vulnerability count. This could be simple as adding more vulnerability scanners, changing the scanning schedules, or determining which priorities need change. Adding more scanners will ease the bandwidth and congestion from one scanner and distribute the workload amongst added scanners. Changing the schedule allows flexibility in reaching more assets since certain assets are offline after core business hours. Adjusting the priorities will require strategic thinking, such as which vulnerabilities change and how and when to consider remediation.

The concept of risk-based thinking and adaptive security principles has been used quite frequently in technology devices. The concepts mirror and provide identical processes that are transferrable. Each addresses the learned behavior patterns and how to prepare, predict, and respond with analytics. It's probably never mentioned, but the adaptive mindset has to analyze facts, metrics, current events, and risk priorities to sustain availability. Various infrastructure devices perform and operate in the same capacity. Many machines work in today's market and use learned behavioral patterns. They utilize the traffic analysis concept to decipher and profile behavioral patterns, which detect network patterns. When the signature profile is created, it stores the baseline information and traffic patterns. The results are compared to determine deviations and anomalies. The adaptive mindset mirrors the same procedure.

When managers and their team members learn the normal system behaviors, they store the information as a mental note. They can identify deviations during adverse events or changes and rapidly counter any security breach, attack, or illegal action. The skillset allows forward-thinking and innovative ideas to prioritize security. In the CISSP domain, the outcome would resemble access control fundamentals or the human firewall theory. With the adaptive mindset, a person restricts security failures from accessing the network, which mirrors a firewall. As the changes occur, they compare the current events and mental notes. If there is a deviation or anomaly, they execute their defense strategy and engage their "adaptive mindset."

The adaptive mindset's sole engagement means that the organization or personnel has adopted security principles. It also provides evidence of a transition from traditional thinking to accepting newer management strategies, technology innovation, and enterprise protective schemes. Last, as the landscape evolved, they also adapted and operated through normalcy. Some of the traditional methods, such as manual operations, transitioned into automated service.

Typical service offerings such as artificial intelligence (AI), big data analytics, and Internet of Things (IoT) replaced manual operations. For instance, the AI effect allowed the business to gain prompt notice and make informed decisions based on intelligence, behaviors, and trends. Its automated detection and response processes reduce costs for human-driven intervention and detection processes. Many threat logs carry about five million events, so could the manual procedure identify all the hostile events? Despite the best tools utilized, teams failed to identify malicious events where security automation surfaced. When deploying the adaptive mindset, one gains information and understands that leading-edge technology can safeguard and provide early warnings. The point is that the adaptive mindset is a growing and evolving practice. When transforming security, every involved unit or personnel must grow their adaptive mindset. The outcome leverages the opportunity to improve enterprise protection, safeguard critical data, defend against intrusions, and adapt and respond to changing operational requirements.

The operational changes are well involved with digital modernization. As corporations research innovative technologies and face different threats, they must adjust the personnel issue. People will only adapt when they foresee a winning edge. The term "winning edge" means that digital modernization is successful and adds value to their working environment. In a traditional discussion, the adaptive mindset focuses on resilience and technology changes, but lurking amongst the ideology are people. How does a business transform the human intellect to adopt newer practices and a solid but agile adaptive mindset? It starts early rather than later within the business model. As companies develop, they must define personnel as assets that empower and protect the enterprise. When security events occur, teams must collaborate and carry out the security objectives. When excluding personnel, the outcome defeats digital modernization, as it's a changing practice that identifies people as assets. The goal is to build a cyber force that is willing and capable of defending enterprises. It's significant since cyberspace contains emerging threats and workforce demands that are escalating. Cyberspace is an ecosystem that requires constant attention and regular observation. Having a functional but adaptive mindset can counter many threats and enable the best talent. All too often, security discussions focus on technology and protection schemes and not skill. Through digital modernization, the issue is well resolved. It articulates people as a concerning force and vital assets. When they are excluded, an adaptive mindset is required. It's a functional state that understands the changing environment and reinforces humans as assets versus a liability. When the concept becomes active, we are now efficient in defending against very intrusive attacks, since the incident response program is a people process.

The adaptive mindset and intrusions are well connected. As attacks surface and environments face different risks, the intelligence and data-sharing initiative must be successful. Each day hackers gain information and share critical steps to infiltrate systems or networks. When examining the adaptive mindset, the technology industry has to, in some form, adapt and make necessary enforcements and acquire knowledge and intelligence that's adaptable. Naturally, hackers operate by different rules and games, and entities that adopt other security practices enable advance protection and warnings. Hackers are abreast of additional counterattack capabilities and utilize the intelligence to design intelligent intrusions and malware programs. Sometimes the secret sauce must change and adapt to enemy tactics. In the combat warfare space, many units gain an informational advantage and conceal their tactical engagement strategies. These strategies could be simple: they contain critical information that is no longer useful or keep their movement and surveillance hidden. The adaptability and skillsets must also operate in a similar mindset. As threats and hackers become more vigorous and their capabilities expand, entities must adopt critical capabilities that can "outthink" their plan. The concept requires the most accurate intelligence and an adaptive mindset that articulates counteroffensive opportunities and intelligent responses.

The Cyber Threat Intelligence (CTI) process is the core and enabler for the adaptive mindset. In today's CTI environment, hackers are utilizing AI to determine events and business behavioral patterns. When data is transmitted across various segments, their AI tool detects the time, type, and recipient. Hackers operate along the same lines because they can adapt their strategies and design sophisticated tools and programs. The defending communities must always do the same but from an ethical perspective. The tools must model the hacker tactics and defend against their malware programs or other deceptive methods. That's why research, intelligence, and information sharing is a vital tool. All three provide the necessary strength to counter hackers' techniques.

To counter and adapt to a changing environment also requires accurate but valuable data. Various programs and processes feed the adaptive mindset; lately, machine learning (ML) has been a practice tool. Since the security landscape, hackers' tactics, and threats constantly evolve, it's imperative to use multi-defense models or concepts. ML can provide a more rapid and adaptive approach to countering intrusions and attacks. As stated, the core of the adaptive mindset is intelligence gathering. The ML approach can rapidly correlate data as it's a subset of AI. ML operates at the root level and utilizes computer algorithms to pinpoint anomalies that humans cannot find. AI imitates human behaviors such as problem-solving, learning, and planning. What's unique is that both replace standard procedures in preventing cyber-attacks. Artificial intelligence can automate complex processes and rapidly detect and react to attacks.

ML enhances AI by delivering measurable results in a structured format that rapidly anticipates and detects attacks and later invokes data analysis techniques that prevent aggression. Both technologies enhance the adaptive mindset intelligence and keep organizations well informed concerning hackers' tactics and strategies. When history is examined, we can foresee that there have been many attacks that occurred via viruses, worms, trojan horses, spyware, or adware. AI and ML in today's market are well suited to help counter and build the adaptive mindset—even when data changes. Sometimes our perspective needs that external support, which is AI and ML. When their data patterns change, many entities may invoke observational practices or continuous operations. They both help to foresee what cyber adversaries utilize data patterns or behaviors, whether state-sponsored, criminal, terrorist, or other unethical practices. The outcome shapes the adaptive mindset through intelligence and accurate data—this is where the "cybersecurity thinking mode" gains success.

The constant observation and intelligence gathering is an ongoing effort. As cyber professionals, it's the norm and acceptable to utilize ongoing compliance to determine the change and security state. NIST has created a valuable Information Security Continuous Monitoring (ISCM) tool to help with security monitoring. It is heavily utilized through technology to determine whether various security objectives, controls, and practices comply. The outcome provides risk indications and trusts that the defense model is being upheld.

When aligning the adaptive mindset and ISCM, entities can utilize situational awareness to determine deviations and required expectations. They can accurately manage risks and prevent attacks through assessments, analysis, and diagnosis of the cybersecurity posture and operational readiness. This is very important since organizational assets, threats, and vulnerabilities are ever-changing.

The 360-degree security visibility principle must be established when deploying the adaptive mindset, as ISCM is its enabler and primary source. The outcome allows entities to shift and remain resilient since ISCM sustains security controls, operations, and stated practices. The inclusion of the adaptive mindset combined with situational awareness mitigate deviations, threats, configuration changes, or uphold baseline operations for the ISCM model. At the core, this is what drives adaptation. Changes will occur, but constant and continuous engagement can rapidly adapt and invoke the necessary remediation or mitigation tactic. In some sense, ISCM is the encapsulated practice that fulfills the adaptive mindset security principles. Every administrative, operational, or technical engagement with the adaptive mindset is combined within the ISCM model. It allows entities to respond since they have prior data intelligence concerning normal operations and the security baseline.

The adaptive mindset is a standard thinking mode. When utilizing cybersecurity, there must always exist "thinking" that shapes the entire process. This is well documented within the adaptative principles, as it's the thinking mode of operations. In cyberspace or its ecosystem, corporations encounter many challenges. They can arrange from management practices, hackers' intelligence, changing events, or intrusions. Despite the obstacles, remaining vigilant and resisting failure provides defense and upholds the holistic defense model. Every expert knows that adapting to change is reasonable and requires practice—not late but early. The "Cybersecurity Thinking Mode" is a state of readiness. It's where organizations fulfill and foresee the protective measures, defense strategies, and threats as everyday concerns. Lapsing or removing elements from the defense model can be risky, so invoking the thinking mode is an ongoing and continuous practice.

The entire discussion for ISCM and adaptive mindset was designed to be last. The scheme is to demonstrate the knowledge for the adaptive mindset and then align the concept to ISCM. These are the "cybersecurity pearls"—as they are where the "thinking mode" survives and generates an ongoing and functional defense practice. The everyday interactions and business relations will utilize the concepts as practical tools that shape security engagements and thoughts. Earlier, the discussion toward readiness and availability was introduced; and by far, they are ongoing and continuous. Establishing a security state is not an easy task, especially when the availability is challenged. Onboarding the cybersecurity pearls can ease the burden and articulate readiness. Despite which security team, unit, or operational demand, the readiness condition is most important. Without its inclusive and branded image, it's impossible to drive the virtual path and transform cybersecurity—so invoking a thinking mode can resolve the weakness and demonstrate how, what, and when to adapt. Last, no person has all the answers, but having a thinking process enables a step forward. Keep thinking and deploy the Cybersecurity Mindset!

CONCLUSION

It has been a great experience to deliver a much-needed resource for the technical and non-technical communities. Every day, we face a myriad of challenges in protecting enterprises and grasping the cybersecurity approach. The Cybersecurity Mindset is a thinking process and functional usage that can harness security and defend systems. Throughout the chapters, you have gained an informational advantage that will sharpen your security engagements and thoughts. When researching the book, various information outlets provided newer or enhanced data for *The Cybersecurity Mindset*. As a reader, I challenge you the same. Continue learning and utilize the reading as a best practice and reference. If you are a college student, CISO, or exploring a career change or just curious about cybersecurity, *The Cybersecurity Mindset* is the knowledgeable resource you need in your arsenal. It prepares your mindset for the cybersecurity journey and serves as a daily read or learning vehicle.

The Cybersecurity Mindset is not an all-inclusive solution since the world is ever-changing. It's designed to provide the initial engagement and thinking process to build our thoughts, and when this happens, we can develop advanced solutions. As the industry matures, so should career milestones and the mindset. Every day our philosophy dictates the decisions, actions, and responses required to sustain security. *The Cybersecurity Mindset* has presented many of those factors, and now, you are mentally armed to defend enterprise assets and understand the cybersecurity blueprint. Since technology changes daily, there is no clear picture about cyberspace, but our thinking mode can advance security. We are all cyber-thinkers and must remain vigilant and use the Cybersecurity Mindset as a common practice. Hackers are getting wiser—so why not our communities as well!

THANKS FOR READING
THE CYBERSECURITY MINDSET

APPENDIX A

ACRONYMS

ADDIE	Analyze, Design, Develop, Implement, and Evaluate
AFB	Air Force Base
AFCEA	Armed Forces Communication Electronics Association
AI	Artificial Intelligence
AWS	Amazon Web Service
B2B	Business to Business
CA	Continuous Assessment
CAP	Corrective Action Plan
CDC	Center for Disease Control
CEH	Certified Ethical Hacker
CERT	Computer Emergency Readiness Team
CIA	Confidentiality, Integrity, and Availability
CIC	Combat Information Center
CISA	Cybersecurity and Infrastructure Security Agency
CISCP	Cyber Information Sharing and Collaboration Program

CISO	Chief Information Security Officer
CISSP	Certified Information System Security Professional
CM	Configuration Management
CND	Computer Network Defense
CNL	Center for Naval Leadership
COBOL	Common Business-Oriented Language
COTS	Commercial Off-the-Shelf Software
COVID-19	Coronavirus Disease 2019
COVID	Coronavirus Disease
CRSC	Computer Security Resource Center
CSP	Cloud Service Provider
CSSLP	Certified Secure Software Lifecycle Professional
CVE	Common Vulnerabilities and Exposures
DHS	Department of Homeland Security
DIACAP	DoD Information Assurance Certification and Accreditation Process
DISA	Defense Information Security Agency
DNA	Deoxyribonucleic Acid
DNS	Domain Name System
DOS	Disk Operating System
DQM	Data Quality Management
EHR	Electronic Health Record
EVP	Employee Value Proposition
FC	Fire Controlman
FedRAMP	Federal Risk and Authorization Management Program
FISMA	Federal Information Security Management Act
FORTRAN	Formula Translation
GIAC	Global Information Assurance Certification
GRC	Governance Risk and Compliance
HBSS	Host-Based System Security
HIPAA	Health Insurance Portability and Accountability Act

HITRUST	Health Information Trust Alliance
HR	Human Resource
HVA	High Value Asset
HWM	High-Water Mark
IA	Information Assurance
IATF	Information Assurance Technology Framework
IBM	International Business Machines
INFOCON	Information Operations Condition
IR	Incident Response
IRP	Incident Response Plan
ISACA	Information Systems Audit and Control Association
ISCM	Information Security Continuous Monitoring
ISFSA	Information Sharing for Situational Awareness
ISO	International Organization for Standardization.
IT	Information Technology
KPI	Key Performance Indicators
KRI	Key Risk Indicators
LAN	Local Area Network
LPOLC	Leading Petty Officer Leadership Course
MIM	Man in the Middle
ML	Machine Learning
NDA	Non-Disclosure Agreement
NETBIOS	Network Basic Input/Output System
NFL	National Football League
NICCS	National Initiative for Cybersecurity Careers and Studies
NICE	National Initiative for Cybersecurity Education
NIST	National Institute of Standards and Technology
NLDP	Naval Leadership Development Plan
OIG	Office of Inspector General
OJT	On-the-Job Training
OMB	Office of Management and Budget

OS	Operating System
OSINT	Open-Source Intelligence
PC	Personal Computer
PHI	Protected Health Information
PII	Personal Identifiable Information
PO	Petty Officer
POA&M	Plan of Action & Milestone
RD&O	Risk Discovery and Opportunities
RMF	Risk Management Framework
ROI	Return on Investment
SA	Situational Awareness
SAAF	Situational Awareness Assessment Framework
SDLC	System Development Lifecycle
SIA	Security Impact Analysis
SIEM	Security Information and Event Management
SLA	Service Level Agreement
SME	Subject Matter Expert
SOC	Security Operational Center
SOW	Statement of Work
SQL	Structured Query Language
SSP	System Security Plan
STIG	Security Technical Information Guide
VM	Vulnerability Management
VPM	Value Proposition Mentality
VPN	Virtual Private Network
VR&P	Vulnerability Remediation and Planning

APPENDIX B

KEY TERMS

Access—Ability and means to communicate with or otherwise interact with a system, use system resources to handle information, gain knowledge of the information the system contains, or control system components and functions.

Accountability—The security goal generates the requirement for actions of an entity to be traced uniquely to that entity. This supports non-repudiation, deterrence, fault isolation, intrusion detection and prevention, and after-action recovery and legal action.

Active Attack—An attack that alters a system or data.

Advanced Persistent Threats (APT)—An adversary with sophisticated levels of expertise and significant resources allows it to create opportunities to achieve its objectives by using multiple attack vectors (e.g., cyber, physical, and deception). These objectives typically include establishing and extending footholds within the information technology infrastructure of the targeted organizations for purposes of exfiltrating information, undermining or impeding critical aspects of a mission, program, or organization, or positioning itself to carry out these objectives in the future. The advanced persistent threat: (i) pursues its objectives repeatedly over an extended period; (ii) adapts to defenders' efforts to resist it; and (iii) is determined to maintain the level of interaction needed to execute its objectives.

Availability—Ensuring timely and reliable access to and use of information.

Backup—A copy of files and programs made to facilitate recovery, if necessary.

Compliance—Compliance focuses on the kind of data handled and stored by a company and what regulatory requirements (frameworks) apply to its protection.

Computer Incident Response Team (CIRT)—Group of individuals usually consisting of security analysts organized to develop, recommend, and coordinate immediate mitigation actions for containment, eradication, and recovery resulting from computer security incidents. Also called a Computer Security Incident Response Team (CSIRT) or a CIRC (Computer Incident Response Center, Computer Incident Response Capability, or Cyber Incident Response Team).

Computer Security Incident Response Team (CSIRT)—A capability set up to assist in responding to computer security-related incidents, also called a Computer Incident Response Team (CIRT) or a CIRC (Computer Incident Response Center, Computer Incident Response Capability).

Continuous Monitoring—The process implemented to maintain a current security status for one or more information systems or for the entire suite of information systems on which the operational mission of the enterprise depends. The process includes 1) the development of a strategy to regularly evaluate selected IA controls/metrics, 2) recording and evaluating IA relevant events and the effectiveness of the enterprise in dealing with those events, 3) recording changes to IA controls, or changes that affect IA risks, and 4) publishing the current security status to enable information-sharing decisions involving the enterprise.

Culture—A group of people or team that shares the same values and goals. An inclusive culture involves the full and successful integration of diverse people into a workplace or industry that shares a common goal.

Cyber Attack—An attack, via cyberspace, targeting an enterprise's use of cyberspace to disrupt, disable, destroy, or maliciously control a computing environment/infrastructure or destroy the integrity of the data or stealing controlled information.

Cyber Incident—Actions are taken using computer networks that result in an actual or potentially adverse effect on an information system and the information residing therein. *See Incident.*

Cyber Infrastructure—Includes electronic information and communications systems and services and the information in these systems and services. Information and communications systems and services are composed of all hardware and software that process, store, communicate information, or combine all these elements. Processing includes the creation, access, modification, and destruction of data. Storage includes paper, magnetic, electronic, and all other media types. Communications include sharing and distribution of information. For example, computer systems; control systems (e.g., supervisory control and data acquisition SCADA); networks, such as the Internet; and cyber services (e.g., managed security services) are part of cyberinfrastructure.

Cybersecurity—The ability to protect or defend the use of cyberspace from cyber attacks.

Cyberspace—A global domain within the information environment consisting of the interdependent network of information systems infrastructures including the internet, telecommunications networks, computer systems, and embedded processors and controllers.

Federal Information Systems Security Educators' Association (FISSEA)—An organization whose members come from federal agencies, industry, and academic institutions devoted to improving the IT security awareness and knowledge within the federal government and its related external workforce.

High Availability—A failover feature to ensure availability during device or component interruptions.

High Impact—The loss of confidentiality, integrity, or availability that could be expected to have a severe or catastrophic adverse effect on organizational operations, organizational assets, individuals, other organizations, or the national security interests of the United States; (i.e., 1) causes a severe degradation in mission capability to an extent and duration that the organization can perform its primary functions, but the effectiveness of the functions is significantly reduced; 2) results in significant damage to organizational assets; 3) results in significant financial loss, or 4) results in severe or catastrophic harm to individuals involving loss of life or severe life-threatening injuries).

Incident—A violation or imminent threat of a breach of computer security policies, acceptable use policies, or standard security practices.

Incident Handling—The mitigation of violations of security policies and recommended practices.

Incident Response Plan—The documentation of a predetermined set of instructions or procedures to detect, respond to, and limit consequences of malicious cyber-attacks against an organization's information system(s).

Information Assurance Vulnerability Alert (IAVA)—Notification generated when an Information Assurance vulnerability may result in an immediate and potentially severe threat to DoD systems and information; this alert requires corrective action because of the severity of the vulnerability risk.

Information Security Continuous Monitoring (ISCM)— Maintaining ongoing awareness of information security, vulnerabilities, and threats to support organizational risk management decisions.

Information Security Continuous Monitoring (ISCM) Program—A program established to collect information following pre-established metrics, utilizing information readily available in part through implemented security controls.

Information Security Risk—The risk to organizational operations (including mission, functions, image, reputation), corporate assets, individuals, other organizations, and the nation due to the potential for unauthorized access, use, disclosure, disruption, modification, or destruction of information and information systems. *See Risk.*

Key Performance Indicators—Key Performance Indicators measure how well something is being done; KPI measures the performance of a specific activity at a predetermined level or amount within a particular amount of time. When metrics reflect the achievement of the desired state, they become KPIs.

Key Risk Indicators—Key Risk Indicators are an early warning to identify a potential event or exposure that may harm the continuity of the activity, project, or mission. When metrics provide an earlier notice regarding an increased risk exposure in a specific area of operations, they become Key Risk Indicators (KRIs).

Low Impact—The loss of confidentiality, integrity, or availability that could be expected to have a limited adverse effect on organizational operations, organizational assets, individuals, other organizations, or the national security interests of the United States; (i.e., 1) causes a degradation in mission capability to an extent and duration that the organization can perform its primary functions, but the effectiveness of the functions is noticeably reduced; 2) results in minor damage to organizational assets; 3) results in minor financial loss, or 4) results in minor harm to individuals).

Mindset—A way of thinking. In a word, a mental inclination or disposition, or a frame of mind. The collection of thoughts and beliefs that shape habits and behaviors.

Moderate Impact—The loss of confidentiality, integrity, or availability that could be expected to have a severe adverse effect on organizational operations, organizational assets, individuals, other organizations, or the national security interests of the United States; (i.e., 1) causes significant degradation in mission capability to an extent and duration that the organization can perform its primary functions, but the effectiveness of the functions is significantly reduced; 2) results in significant damage to organizational assets; 3) results in significant financial loss, or 4) results in significant harm to individuals that do not involve loss of life or severe life-threatening injuries).

Risk—The level of impact on organizational operations (including mission, functions, image, or reputation), corporate assets, or individuals results from the process of an information system given the potential impact of a threat and the likelihood of that threat occurring.

Risk Analysis—The process of identifying the risks to system security and determining the likelihood of occurrence, the resulting impact, and the additional safeguards that mitigate this impact. Part of risk management and synonymous with risk assessment.

Risk Assessment—The process of identifying risks to organizational operations (including mission, functions, image, or reputation), corporate assets, individuals, other organizations, and the nation, arising through the process of an information system.

Risk Assessment Methodology—A risk assessment process, together with a risk model, assessment approach, and analysis approach.

Risk Assessor—The individual, group, or organization responsible for conducting a risk assessment.

Risk Management—The process of managing risks to organizational operations (including mission, functions, image, reputation), corporate assets, individuals, other organizations, and the nation, resulting from the process of an information system and includes: (i) the conduct of a risk assessment; (ii) the implementation of a risk mitigation strategy; and (iii) employment of techniques and procedures for the continuous monitoring of the security state of the information system.

Risk Management Framework—A structured approach is used to oversee and manage risk for an enterprise.

Risk Mitigation—Risk mitigation refers to the process of planning and developing methods and options to reduce threats—or risks—to project objectives.

Risk Monitoring—Maintaining ongoing awareness of an organization's risk environment, risk management program, and associated activities to support risk decisions.

Risk Response—Accepting, avoiding, mitigating, sharing, or transferring risk to organizational operations (i.e., mission, functions, image, or reputation), corporate assets, individuals, other organizations, or the nation.

Security Information and Event Management (SIEM) Tool—An application that provides the ability to gather security data from information system components and present that data as actionable information via a single interface.

Situational Awareness—Knowing an environment and knowledge to predict and respond based on senses, information, and previous encounters. It involves drawing a kind of mental map that helps understand what is happening.

Social Engineering—An attempt to trick someone into revealing information (e.g., a password) can be used to attack systems or networks.

Threat—Any circumstance or event with the potential to adversely impact organizational operations (including mission, functions, image, or reputation), corporate assets, individuals, other organizations, or the nation through an information system via unauthorized access, destruction, disclosure, modification of information, and denial of service.

Threat Event—An event or situation that has the potential for causing undesirable consequences or impact.

Threat Monitoring—Analysis, assessment, and review of audit trails and other information collected to search out system events that may constitute violations of system security.

Threat Source—The intent and method targeted at the intentional exploitation of a vulnerability or a situation and strategy may accidentally trigger a vulnerability. Synonymous with Threat Agent.

User—Individual or (system) process authorized to access an information system.

User ID—Unique symbol or character string used by an information system to identify a specific user.

Value Proposition—A business or marketing statement summarizes why a consumer should buy a product or use a service. This statement should convince a potential consumer that one product or service will add more value or better solve a problem than other similar offerings.

Virtual Private Network (VPN)—A virtual network built on top of existing physical networks provides a secure communications tunnel for data and other information transmitted between networks.

REFERENCES

Bank of England (2016). *CBEST intelligence-led testing. Understanding cyber threat intelligence operations,* Version 2.0. Retrieved January 25, 2021 from https://www.bankofengland.co.uk/-/media/boe/files/financial-stability/financial-sector-continuity/understanding-cyber-threat-intelligence-operations.pdf

Capon, N., & Hulbert, J. (2008). *Managing marketing in the 21st century: Developing & implementing the market strategy.* Wessex, Inc. Retrieved December 22, 2020 from https://www.amazon.com/Managing-Marketing-21st-Century-Implementing/dp/0979734401

Carnegie Mellon University (2016). *CRR supplemental resource guide—Vulnerability management.* 4(1.1), 11-12. Retrieved August 23, 2020 and February 15, 2021 from https://www.cisa.gov/sites/default/files/publications/CRR_Resource_Guide-VM_0.pdf

Chron Contributor (2020). *Three types of interdependence in an organizational structure.* Retrieved February 12, 2020 from https://smallbusiness.chron.com/three-types-interdependence-organizational-structure-1764.html

CISCO (2017). *The essential guide to workplace modernization through next-generation unified communications. The new face of the modern workplace.* Retrieved December 19, 2020 from https://www.oneneck.com/hubfs/Blog/Blog%20Downloads/ Essential%20Guide%20to%20Workplace%20Modernization.pdf

College of the Holy Cross (2017*). Policies and Procedures Manual: Written information security plan,* Policy #: 1-0100 (Version 1.3). IT Policy & Security Committee. Retrieved April 18, 2021 from https://www.holycross.edu/sites/default/files/files/its/pdf/1- 0100_written_information_security_plan_1.3.pdf

Dutton, W. H. (2017). Fostering a cyber security mindset internet policy review, 6(1), 5-6. *Internet Policy Review.* Retrieved January 23, 2020 from https://doi.org/10.14763/2017.1.443

Dweck, C. S. (2017). *Mindset: Changing the way you think to fulfil your potential* (6th ed.). https://www.amazon.com/Mindset- Updated-Changing-Fulfil-Potential/dp/147213995X/ref=sr_1_ 2?dchild=1&qid=1612909685&refinements=p_27%3ADr+Car ol+Dweck&s=books&sr=1-2

Glessner, D. (2019). *Digital transformation: More about mindset than technology.* Genpact Blog. Retrieved December 19, 2020 from https://www.genpact.com/insight/blog/digital-transformation- more-about-mindset-than-technology

Hassan, A. (2012). The value proposition concept in marketing: How customers perceive the value delivered by firms—A study of customer perspectives on supermarkets in Southampton in the United Kingdom. *International Journal of Marketing Studies, 4*(3), 69-70. Retrieved January 23, 2021 from https:// doi.org/10.5539/ijms.v4n3p68

Hoftstadter, D. R. (1999). *Gödel, Escher, Bach: An eternal golden braid* (20th ed.). Basic Books. Retrieved February 5, 2020 from https://www.amazon.com/G%C3%B6del-Escher-Bach-Eternal-Golden/dp/0465026567/ref=sr_1_1?dchild=1&keywords=ISBN +9780465026852&qid=1614643497&sr=8-1

HSE (2012). *Leadership and worker involvement toolkit: Knowing what is going on around you (situational awareness)—Information sheet from seven steps, step 6, key tool and further tools.* Retrieved August 23, 2020 from https://www.hse.gov.uk/construction/ lwit/assets/downloads/situational-awareness.pdf

ISACA (n.d.). *The cybersecurity culture gap. An ISACS and CMMI Institute study.* Retrieved February 20, 2020 from https://www. isaca.org/-/media/info/cybersecurity-culture-report/index.html

Kim, P. (2014). *The hacker playbook: Practical guide to penetration testing.* CreateSpace Independent Publishing Platform. Retrieved April 2, 2020 from https://www.amazon.com/Hacker-Playbook-Practical-Penetration-Testing/dp/1494932636/ref=sr_1_1?dch ild=1&keywords=ISBN+978-1494932633&qid=1614637959& sr=8-1

Mindset Works, Inc. (2017). *Dr. Dweck's research into growth mindset changed education forever.* Retrieved March 24, 2020 from https://www.mindsetworks.com/science/

Murray, S. A., Ensign, W., & Yanagi, M. (2010). *Combat situation awareness (CSA): Model-based characterizations of Marine Corps training and operations* (SPAWAR Technical Report 1994). Retrieved April 14, 2021 from https://apps.dtic.mil/dtic/ tr/fulltext/u2/a535225.pdf

National Initiative for Cybersecurity Careers and Studies (2021). *Workforce framework for cybersecurity: NICE framework.* Retrieved January 19, 2021 from https://niccs.cisa.gov/workforce-development/cyber-security-workforce-framework

National Institute of Standards and Technology (2012). *Guide for conducting risk assessments: Information security* (NIST special publication 800-30, Revision 1). U.S. Department of Commerce. Retrieved September 10, 2020 from https://nvlpubs.nist.gov/nistpubs/Legacy/SP/nistspecialpublication800-30r1.pdf

National Institute of Standards and Technology (2020). *Risk management framework for information systems and organizations: A system life cycle approach for security and privacy* (NIST special publication 800-37, Revision 2). U.S. Department of Commerce. Retrieved September 10, 2020 from https://nvlpubs.nist.gov/nistpubs/SpecialPublications/NIST.SP.800-37r2.pdf

National Institute of Standards Technology (2020). *Security and privacy controls for federal information systems and organizations* (NIST special publication 800-53, Revision 5). U.S. Department of Commerce. Retrieved September 10, 2020 from https://nvlpubs.nist.gov/nistpubs/SpecialPublications/NIST.SP.800-53r4.pdf

Office of Information Technology (2018). *Cloud strategy roadmap, FY18 & FY19.* Office of Information System Management. U.S. Department of Veterans Affairs. Retrieved July 15, 2020 from http://vistaadaptivemaintenance.info/va-cloud/VA_Cloud_Strategy-Roadmap-2018-2019.pdf

Office of Prepublication and Security Review (2019). *DoD digital modernization strategy. DoD information resource management strategic plan*, FY 19-23. U.S. Department of Defense. Retrieved December 22, 2020 from https://media. defense.gov/2019/Jul/12/2002156622/-1/-1/1/DOD-DIGITAL-MODERNIZATION-STRATEGY-2019.PDF

O'Neil, D. (2006). *Human culture: What is culture?* Retrieved February 13, 2020 from https://www2.palomar.edu/anthro/ culture/culture_1.htm

Singer, P. W., & Friedman, A. (2014). *Cybersecurity and cyberwar: What everyone needs to know* (1st ed.). Oxford University Press. Retrieved January 25, 2021 from https://www.amazon.com/ Cybersecurity-Cyberwar-Everyone-Needs-Know%C2%AE/ dp/0199918112/ref=sr_1_1?dchild=1&keywords=ISBN+978-0199918119&qid=1614635314&sr=8-1

Stratfor (2012). *On security: A practical guide to situational awareness*. Retrieved August 23, 2020 from https://worldview.stratfor.com/ article/practical-guide-situational-awareness

Taylor, P. A. (1999). *Hackers: Crime in the digital sublime*. Routledge. Retrieved February 2, 2021 from https://www.amazon.com/ Hackers-Digital-Sublime-Paul-Taylor/dp/0415180724/re f=sr_1_1?dchild=1&keywords=ISBN+978-0-415-18072-6&qid=1614641510&sr=8-1

Thomas, D. (2002). *Hacker culture*. University of Minnesota Press. Retrieved February 2, 2021 from https://www.amazon. com/Hacker-Culture-Douglas-Thomas/dp/0816633452/ ref=sr_1_1?dchild=1&keywords=.+ISBN+0-8166-3345-2&qid=1614641705&sr=8-1

U.S. Army (2019). *Army modernization strategy: Investing in the future*. Retrieved September 10, 2020 from https://www.army.mil/e2/downloads/rv7/2019_army_modernization_strategy_final.pdf

U.S. Department of Homeland Security (2018). *Cybersecurity strategy*. Retrieved October 22, 2020 from https://www.dhs.gov/sites/default/files/publications/DHS-Cybersecurity-Strategy_1.pdf

U.S. Department of Homeland Security (2019). *Privacy impact assessment for the National Cybersecurity Protection System (NCPS)—Intrusion detection* DHS/CISA/PIA-033. Retrieved December 22, 2020 from https://www.dhs.gov/sites/default/files/publications/privacy-pia-dhscisa033-ncpsintrusiondetection-sept2019.pdf

Website Builder (2021). *30 Key cybersecurity statistics to be aware of in 2021*. Retrieved April 14, 2021 from https://websitebuilder.org/blog/cyber-security-statistics/

Whitehouse (2000). *Management of federal information resource*: *Memorandum for heads of executive departments and agencies* subject (CIRCULAR NO. A-130, Revised Transmittal Memorandum No. 4). Retrieved October 21, 2020 from https://www.whitehouse.gov/sites/whitehouse.gov/files/omb/circulars/A130/a130trans4.pdf

Wu, C. H. J., & Irwin, J. D. (2013). *Introduction to computer networks and cybersecurity* (1st ed.). CRC Press. Retrieved August 23, 2020 from https://www.amazon.com/Introduction-Computer-Networks-Cybersecurity-Chwan-Hwa/dp/1466572132/ref=sr_1_1?dchild=1&keywords=.+ISBN+978-1466572133&qid=1614637707&sr=8-1

Conceptual Model for the Cybersecurity Mindset

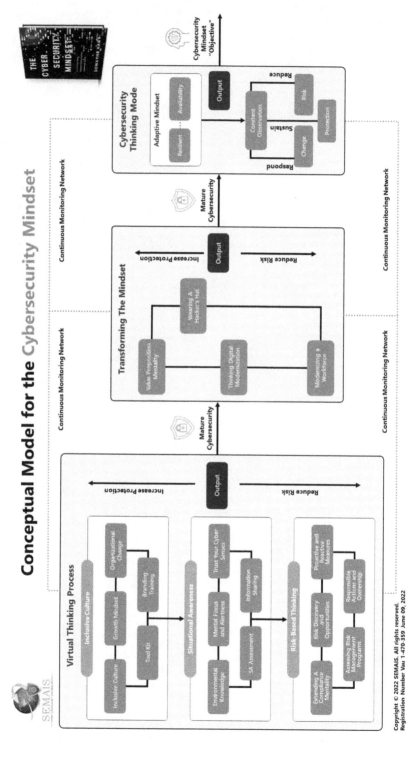